KISHTWAR'S POWER QUEST

Resonia's journey through Kishtwar's power challenges

RESONIA
PANKAJ KUMAR

notionpress.com
INDIA • SINGAPORE • MALAYSIA

ISBN
Paperback 979-8-90296-766-8
Hardcase 979-8-90451-358-0

Contents

Preface

Every large transmission project begins with a purpose.

In Kishtwar, that purpose was direct: to create a system that could carry power across distance, terrain, and season, and continue doing so year after year.

It follows the work as it unfolded on ground, shaped by geography, access, weather patterns, coordination demands, and the daily judgement required to keep progress moving. The narrative stays close to execution, where drawings meet soil, schedules meet roads, and decisions meet consequence.

The project brought together many strands of work. A compact GIS substation arranged across levels on a hillside. A short transmission link tied into an existing 400 kV corridor. Towers placed with care along slopes that allowed little margin. Methods adjusted when equipment access reached its limit. Commissioning carried out with attention fixed on every reading and response. Each stage demanded focus, patience, and alignment across teams.

Beyond engineering, the work depended on people and systems operating in step. Site teams balancing productivity with safety. Planners tracking progress against conditions that changed without warning. Administrators maintaining order across movement, payments, records, and supply. Local dialogue shaping access and continuity. Families supporting long stretches of work carried out far from routine comforts.

The chapters move in sequence, reflecting how the project progressed. Early sections address intent, planning, and design shaped by terrain. Later sections follow base setup, access challenges, execution across monsoon and winter, coordination under security constraints, material flow, financial control, and the transition into energisation. The closing chapters reflect on what the node enabled for the region and what it required from those who built it.

This book carries many voices. Engineers, supervisors, contractors, planners, and support teams appear through their actions and decisions. Their work forms the foundation of what is recorded here. The intent remains consistent throughout: to document how a transmission project advanced through discipline, coordination, and sustained effort.

What follows is a record of work done with care, responsibility, and persistence.

> ***It stands as a reference for how complex infrastructure takes form when people remain attentive to both plan and place, from the first review to the moment power begins to flow.***

Introduction

This is what our team hears every day.
"How are you today."

The first voice speaks before any plan. A guard asks it at the gate. A driver asks it over the radio. A planner asks it on the walk to the yard. The words are small and real. They set a tone that cares for people first.

Work here begins with a look at the ground and the sky. A bend may be fine in the morning and tricky by noon. Frost may sit where it did not yesterday. A small stream may have found a new path. None of this needs drama. A short call goes out. A line on the board changes. A truck takes a safer turn. The day keeps moving because the team moves with sense.

The voices in these pages belong to carpenters, riggers, welders, banksmen, drivers, planners, engineers, guards and many others who stand with them. Families stand close as well. Meal hours shift. School runs adjust. Sleep changes when a crew must leave early or return late. Neighbours see slow trucks in narrow lanes and judge us by what we do, not by what we say. Trust grows when we keep our word in small things.

Power work can look big from far away. Up close it is a string of modest acts, done well and done again. A rope is checked before a lift. A ladder is tied at the proper point. A harness is clipped without hurry. A setting is read twice before a breaker is closed. A record is written in firm letters. None of this makes a headline. All of it keeps harm away and brings people home in one piece.

Hills teach patience. Speed is noise when one tight turn can bring it to a stop. Our crews found a tempo that holds. Tools leave the store with a name and a return time. Fuel runs avoid school traffic. Food reaches early when a pour needs a noon slot. Radios share one channel at set hours. People know when to speak and when to listen. Habits like these look small on paper. On the ground they save time and they save lives.

Planning serves the field, not the other way around. A drawing that works in wind, rain and gloves is a good drawing. A route that lets a truck reverse without risk is a good route. A rest spot under a cover, marked and kept tidy, is a good choice. A plan that respects the bend, the bench and the season survives contact with real work. A wall chart alone has no power. People give it power when it speaks their language.

Five minutes of safety talk can be the best tool in the kit. The lines are short and plain. Wear the helmet. Tie the ladder. Use the tag line. Keep clear of that edge. Mind your hands. Repeat as needed. The same notes play every morning and again after lunch. The floor under the day rises by an inch each time. When tasks turn hard, that floor holds.

Money matters in ways that reach kitchens and school fees. Friday matters in homes that run on weekly wages. Good managers honour that rhythm. Bills match deliveries. Vendors are paid on time. Notices are posted where everyone can read them. A new week starts without quarrels over small slips. It may sound like paperwork. It is respect in action.

Trust outside the fence decides more than any spreadsheet. People who own plots along a route expect us to arrive when we said we would. A police post that keeps an eye out for a late return deserves a thank you in person. A shop that extends credit for ten days deserves payment on the eleventh. These are not lines in a speech. They are visits and handshakes. Work like ours lives on such ground.

The place itself speaks in firm terms. Cut a bench wrong and it will move. Ignore a drain and silt will fill it by evening. Leave a slope without a toe and it will bite back. Crews who learn to read these signs gain skill that travels with them. They start to leave behind work that stands through rain and snow. They learn to value quiet equipment and still walls. Silence in a structure means it is doing its job.

Families sit at the heart of this story. Many entries in a site log trace back to a home. A person who sleeps well works safer. A spouse who trusts a promised return time waits with less fear. A child who studies under a lamp that stays on keeps faith with the lesson. Steel and wires do the carrying. People give the purpose. When evenings hold steady, parents breathe easier. That breath is worth more than any ribbon.

Language here stays close to daily speech. Teams in rough places talk in short words and straight lines because those words do not fail under rain and frost. Fancy talk slips. Plain talk grips. A foreman should be able to read any page at six in the morning and use it by seven. If one line helps a crew avoid harm or helps a manager set a better week, then the time spent writing was worth it.

Names matter. A leader held the line when weeks ran long. A programme head kept plan and ground in one view. Engineering minds set a yard that sits right on a hill. Safety leads wrote rules that held when cold bit through gloves. Quality teams tied winter work to checks that last. Line hands walked bends and marked pull-ins that gave the span its script. Erection crews timed lifts so cranes did not wait. Execution leads turned drawings into routes that crews could follow.

HR and admin kept rooms, vendors and papers steady when phones fell silent. Planners made one task meet the next without gaps. Construction leads fixed snags before they ate a day. Finance teams kept Fridays honest so the week ahead could run. Supply chain found hours the road would allow. Land officers met owners and returned when they said they would. GIS and electrical engineers stood by panels in trial rooms. Civil leads helped set benches that held through rain. These roles tell the truth that large work rests on many hands.

There is no lecture here and no trophy case. There are field notes you can use. We marked where time went away and where we found it again. We wrote down what helped and what hurt. We show the worth of small routines, steady calls and fair dealing. The scenes sit close to daily life. A guard pouring tea for a wet crew near midnight. A driver laughing at a bend that once scared him. An inspector nodding once when a joint passes on the first attempt. These are the moments that carry people through a long season.

Power lines and yards serve people who may never learn our names. That is fine. Worth shows up in quiet ways. A living room with light in the evening. A clinic that runs a device through the night without second thoughts. A small shop that can plan hours without guessing. Crews carry that measure in their heads when they tie a knot or stamp a bill. It keeps them straight when a shortcut whispers. It asks them to build so that the next crew, years later, finds a node that still holds.

It asks more than health. It asks for honesty in one breath. It checks weather and mood without fuss. It lets a foreman say whether to push, pause or move to a plan B that already exists. It keeps charts tied to real people. It gives the day a start that respects human limits and human strength. Hold that call as you read. Hear it on the road into town and again on the climb to the yard. Hear it in a control room before a trial. Hear it at a table when two people settle a small dispute with a calm word.

If we keep asking and answering with care, the work will stand the test of season and storm. The greeting will remain fresh, even on the hundredth day. The reply will guide the next hour. The hour will build the week. The week will leave behind something useful. And tomorrow, when a new crew steps through the gate, the first voice will say it again, as steady as a sunrise.

"How are you today."

WHY THIS PROJECT?

"Empowering humanity by addressing the toughest challenges of clean energy delivery."

It's more than a logo. It's what the company was built to do.

At Resonia, the word *power* has never meant only electricity. It's meant continuity. Safety. A sense of control in daily life. Over the years, the company has stepped into projects that weren't easy to pull off, some in places where the land is difficult, others where the approvals take time, and a few where nothing has been built before. What kept the work going was a single idea: that energy should reach everyone, even when the route looks hard.

The team often spoke about the spaces between the wires, the people, the homes, the moments they don't get to see, but which they know are waiting. A bulb is staying lit in a labour room. A call that doesn't drop in the middle of an emergency. A student who can study late without squinting under a flickering tube. These are not inventions. They are ordinary scenes, and that's exactly why they matter.

Resonia isn't in the business of temporary fixes. It takes on work that demands consistency. Long-term solutions. The kind of systems that don't leave people guessing. Power that stays when the weather turns, when the load increases, when there's pressure on the grid.

Behind the designs and the transmission lines, there's always been something simple at the core: people should be able to rely on power as a part of their life, not as a luxury that comes and goes.

When this project came up, the company didn't have to look twice. The need was obvious. The region had been waiting. The team knew this was one of those moments the motto was made for.

This is where the journey to **Kishtwar** began.

Chapter 1

River, Valley, Grid

It curls along the river, climbs into pine and stone, then slips behind cloud.

Corners arrive quicker than you expect. A bridge appears, then another. A hillside opens for a moment, then folds back in. Travel here asks patience from the hands on the wheel and respect from anyone carrying a deadline.

People who live in these hills grew up with that same discipline. Weather decides plans. Distance stretches every errand. Services reach late, sometimes after years of waiting. Electricity, in particular, came with conditions. It could arrive for an hour, then vanish. Homes learned workarounds. Shops learned closing times shaped by bulbs and backup. Families learned how to keep a day running when the basics refused to cooperate.

Over time, the gap stopped being a complaint and became routine. Yet a hope stayed alive, spoken in ordinary ways, carried in small decisions. When power comes, it should remain. When a promise is made, it should hold.

That is the ground Resonia stepped onto when the team first walked Kishtwar.

Transmission can look like metal, towers, and lines from a distance. On site, it feels like assurance. Something you can rely on when rain hits hard, when slopes refuse easy footing, when access tests every plan. The work asks for systems that keep performing after the last truck turns back and the last camp is packed away.

In a place like Kishtwar, power carries a different kind of meaning. A fan running through a warm night. A hospital staying operational after dusk. A teacher using a projector without second thoughts. These scenes feel ordinary in many towns. Here, they speak of relief and continuity, day after day.

The Work Between the Wires

Every project begins with lines on paper. This one began with memory and consequence.

The team had seen what happens when electricity reaches the far end of a long road. The first change shows up fast. The deeper change arrives in the hours that follow. Shops keep shutters up a little longer. Radios play into the evening. Streets stay active after sunset. People stop timing life around daylight.

Kishtwar brought its own set of demands. Terrain that never gives you the same answer twice. Permissions that require patience and presence. Logistics that call for discipline, not noise. These were familiar tests. What made this journey feel different was the responsibility carried into the region.

Engineers, planners, and crews spoke often of the people they would never meet. They still worked with those lives in mind. A decision on a route, a yard layout, a sequence of lifts, a day held back due to weather, each choice had a human outcome attached to it.

Resonia approaches energy delivery with a long view. Systems are built for endurance, for load, for hard seasons, for days when conditions squeeze every margin. People who rely on these networks deserve reliability they can build a life around.

When Kishtwar came up, the need stood in front of everyone involved. A region that had waited. A population that had adapted because there was no other way. A future ready to begin with one decision, followed by months of committed work.

Now, as the lines stretch across valleys and over hills, they carry more than current. They carry continuity. They carry evenings that run their full length. They carry a sense that the day will behave the way it should.

This journey will live beyond reports and photographs. It will remain in homes after dusk. In clinics that keep running through the night. In classrooms where a lesson stays on track. In families who stop checking the switch twice.

These are the outcomes that matter. These are the stories that remain.

Resonia will keep walking into regions where conditions ask patience and skill, and where progress shows itself in daily life. Somewhere between valleys, on a far hillside, someone waits for the light to stay.

That remains reason enough.

Electricity in Jammu & Kashmir has never been easy. In many homes, people are used to planning around power cuts. Some keep a spare torch by the door. Others cook early, just in case. The lights go off without warning, and the wait begins. It has become part of life.

But this wasn't because the region lacked energy sources. The opposite is true.

Jammu, Kashmir, and Ladakh are surrounded by rivers that can generate a large amount of hydropower. These rivers, Chenab, Jhelum, Ravi, and parts of the Indus, carry strong currents across high terrain. This kind of flow is ideal for producing clean electricity through hydroelectric projects. The numbers show what's possible.

The total hydropower potential of the region is estimated at **20,000 megawatts (MW)**. Out of this, **16,475 MW** has already been surveyed and identified. Most of it lies in the Chenab basin (**11,283 MW**), followed by **3,084 MW** in the Jhelum basin, **1,608 MW** in the Indus, and **500 MW** in the Ravi basin. Even with all this identified, only **3,263 MW**, or about **20%**, has been developed and added to the grid.

This gap is important. It means the state has the resources but hasn't yet built enough infrastructure to turn them into reliable supply. As a result, demand keeps growing while electricity remains uneven.

The demand for power in the region is rising fast. Between **2004–05** and **2018–19**, it went up from **1,706 MW** to **4,217 MW**. In terms of actual energy used, this meant a jump from **9,640 million units** to **21,887 million units** within less than two decades. The All India Power Survey noted that this trend will continue.

Meeting this demand isn't about large cities alone. Smaller towns, remote villages, schools, health centres, and local businesses all depend on steady supply. The better the connection, the more they can grow.

The state has enough water to generate power. What it needs is a way to move that power, smoothly, safely, and across seasons. That's where transmission matters. That's where the work begins.

Why Kishtwar Specifically?

Kishtwar is where many of the region's biggest hydropower projects are being built. The Chenab River flows through it with strength and consistency, making the area well-suited for generating electricity. Projects like **Pakal Dul (1,000 MW)**, **Kiru (624 MW)** and **Kwar (540 MW)** are all located nearby. Together, they bring in over **2,100 megawatts** of planned capacity.

This makes Kishtwar more than just another point on the map. It's where power from all three projects can be brought together. Once collected, it can be routed into the grid, helping both the state and the northern region manage their electricity needs more reliably.

The location works well. It sits close to the generation sites and can link directly with larger transmission lines heading out of the valley. Setting up the system here means building fewer new lines and making the most of the natural layout of the land.

> ***Kishtwar made practical sense. It brought the power projects, the river, and the grid within reach of each other. This was the place to begin.***

Why Resonia Stepped Up

Resonia has always worked in places that need more than routine project delivery. Over the years, they've taken on assignments where the odds were tough, where the geography, permissions, or timelines needed more than technical skill. These weren't always easy to explain on paper, but the team understood what it meant to show up and keep going.

The terrain around Kishtwar wasn't unfamiliar. Resonia had already worked across Jammu and Kashmir before, on the NRSS-29 project, which brought power into the Kashmir Valley through one of the most difficult stretches of land in the country. That experience gave them the footing to look at Kishtwar and know what it would take.

This wasn't about technical drawings or contracts. It was about figuring out whether the team could move material through narrow roads, whether labour could be mobilised in high-altitude zones, and whether the coordination between agencies, locals, and officials would hold through the life of the project.

Someone from the team remembered standing near the site early in the planning stage, looking at the hills ahead. *"It wasn't the land that worried us,"* he said, *"it was time. We had to be fast without cutting corners. That was the challenge."*

This project wasn't planned to fix a shortfall and walk away. It was designed to last. Over the next **35 years**, this single line is expected to carry the region's power, hold it steady, and keep it flowing, through every season, every change in demand, every change in load.

The system is built in such a way that if one line faces a fault or disruption, the power can continue through alternate paths. That kind of resilience matters in a place where bad weather and tough access can delay repair teams by hours, sometimes days. When the line is live, it doesn't just serve one part of the state. It connects with the larger grid that powers the rest of North India. What starts in the Chenab basin can now reach far, across districts, across borders.

The direct benefit of this project will be felt first by the people closest to it. In small towns and villages, stable electricity makes a visible difference. Machines in workshops run longer. Shops stay open without worrying about inverters. Children are able to study at home after sunset without interruption. Clinics can store vaccines safely and run equipment when it's needed most. In places where plans are often adjusted depending on when the power returns, a steady supply allows people to plan properly, to commit to their work, to make investments, to move forward.

This sort of access supports more than just comfort. It makes it easier for new services to enter. It brings in banks, small manufacturing units, schools with better facilities. It removes one of the biggest uncertainties that stops progress at the local level.

For Resonia, this project reflects more than a contract delivered. It's part of a longer journey to strengthen the country's energy backbone while making sure that development doesn't leave anyone behind. The commitment they made through their motto, *empowering humanity by addressing the toughest challenges of energy delivery*, was never meant to sit on a brochure. It was always meant to be tested on the ground.

There's a long way to go. There will be challenges ahead. But this is how the work begins, by choosing a hard place, committing to it fully, and building something that can hold its ground for years to come.

What happens next is the story of how that promise is kept.

Chapter 2

From Bid to Breakthrough

A Project in the Cross-hairs

When Resonia first looked at the Chenab basin, they saw more than just fast-moving rivers and high mountains. The Chenab Valley held something practical and needed, an answer to Jammu & Kashmir's ongoing electricity troubles.

The figures made a convincing argument. Official data clearly showed the region had around **20,000 megawatts** of hydropower potential. From that total, engineers had already mapped out **16,475 megawatts** precisely, much of it concentrated around the Chenab river. Yet the state had managed to tap into less than **20%** of what was clearly available. This gap in development had real consequences. People across the region were getting fewer than 100 megawatts against a daily need of around 145 megawatts. Towns and villages endured daily outages, forcing people to live with uncertainty, shops shutting early, families cooking by candlelight, and children studying around battery life.

Resonia recognised early that this was more than a business opportunity; it was a genuine chance to fix something that affected thousands of everyday lives. For them, the Chenab basin scheme presented a direct route to delivering stable, reliable electricity to local communities, while also feeding surplus power into India's national grid.

But good opportunities rarely come without risks. Early feasibility checks revealed several points that needed close attention. Terrain was a major challenge. The region featured hills at angles between 30 and 45 degrees, making conventional construction difficult. The winters brought temperatures plunging to -6°C and below, accompanied by heavy snowfall that could stall work for weeks. Roads were narrow and not built to handle large equipment such as transformers. There were also uncertainties around gaining community trust. Other projects in the valley had faced strong local protests and delays caused by land acquisition disputes.

These initial concerns were openly discussed within the Resonia team before the bidding started. Project planners knew from previous experience that such risks were manageable, but only if they were handled transparently and proactively. By the time the decision to bid was made, everyone involved understood clearly that winning would mean careful planning, smart logistics, and constant dialogue with local people and officials.

Inside the Bid Room

The Resonia team sat down together to figure out exactly how to approach the Chenab Valley project. They were practical people, engineers who knew their job, planners with experience, and finance professionals who could realistically assess costs. Each person knew their area well and came ready to contribute.

They started by gathering facts. The engineers pulled out detailed maps to see the land clearly. They checked slope angles, pinpointed narrow roads, and marked spots where large equipment might have trouble moving. These practical details shaped the way they thought about the entire project. The team noted issues openly, recognising challenges upfront.

Finance experts worked carefully with real numbers. They broke down the costs into clear sections: labour charges, machinery rentals, transport expenses, and other essentials. They avoided rough estimates. Each number came from solid calculations, based on actual conditions and experiences from similar projects.

Logistics specialists looked closely at local weather patterns and seasonal shifts, especially snowfall and winter conditions. They discussed how heavy snow could affect timelines and equipment movements. Plans were adjusted realistically, allowing enough space to handle unexpected events without major delays.

One important idea emerged from these discussions. Instead of crossing whole kishtwar stretching across difficult mountains, Resonia's team chose a simpler, smarter approach. They identified substation land with challenging but manageable slope and terrain near to tapping point and reducing the line length to tapping point of an existing 400 kV transmission line passing nearby. It was a practical way to reduce construction work, limit the number of new towers, and avoid unnecessary land acquisition.

Letter of Intent, Letter of Commitment

It wasn't long before official word came through. Resonia had won the bid for the Chenab Valley project. The news was welcomed warmly by everyone involved. The project had moved from an idea to something real, something they could now act on. The team celebrated briefly, knowing this was the first step in a long journey ahead.

Almost immediately after the award was announced, Resonia set up a dedicated project team structure to handle the project. They acquired the Special Purpose Vehicle named Kishtwar Transmission Ltd from Bid Process Coordinator PFC Consulting Limited. Acquiring this specific entity showed commitment. It gave the team a single, strong platform to manage everything effectively, from finance to operations.

Speaking of finance, the Resonia team structured funding carefully. The project would run for 35 years, and a realistic revenue framework was in place from the very start. Financial planners arranged a sensible combination of equity and loans. The funding was practical and matched the steady, long-term nature of the project itself. For Resonia, getting the finances right at the start meant fewer headaches down the road.

With these key pieces in place, the official award, the dedicated company, and solid funding, the next task was finding the right people to bring the project to life.

Calling the Crew

Recruiting for a project like this was different from regular hiring. Resonia needed people who genuinely understood the challenges of building in mountainous areas. They searched specifically for specialists who had experience with steep terrains, difficult weather, complicated logistics, and critical security matters. Each team member had to be practical and ready to handle unexpected issues.

Late one night, a call went out to Abhishek Garg. At the time, Abhishek was comfortable working elsewhere. He had no immediate plans to change jobs. The late-night call surprised him. It was Anushul Tripathi, Resonia HR representative, who explained the job was based near Jammu, deep in the hills of Kishtwar.

He took a moment to think about it. He'd spent many years working in flat landscapes across Gujarat, Rajasthan, MP, Bihar, and Haryana. This would be different. The idea of working in hilly regions caught his attention. He had never built a substation before; his experience had mainly been with transmission lines. Abhishek felt this was an opportunity to learn something new. Managing stakeholders in a tough environment also appealed to him. The practical challenge drew him in.

He agreed to consider the offer. Within a week, he went through four rounds of interviews. After those meetings, Abhishek felt sure about joining. For him, it was more than a job change or a higher paycheck. It was a genuine chance to grow personally and professionally by taking on a different kind of challenge.

First Look at the Land

When Abhishek and his team reached Kishtwar for the first time, the ground corrected every assumption they had carried in.

Photos had flattened the hills. On site, they rose hard and fast, slopes climbing at 30 to 45 degrees, cutting into the sky with no easy breaks. Abhishek stood for a moment and looked up, following the line of the ascent, already measuring what it would take to move men and machines across it. The climb sat there like a warning and a challenge in the same breath.

Then there were the roads.

They had been carved decades earlier for small vehicles, for everyday cars and light trucks, long before anyone spoke of hauling transformers and heavy assemblies into this terrain. The route curled tightly, one bend feeding into the next, with stretches where two vehicles could not pass. On the outside edge, the land fell away in a drop that made you keep your foot steady and your eyes sharper.

Abhishek and the engineers stood by the roadside, tracing each turn with their gaze, replaying the route in their heads like a slow rehearsal. They could already see the movement that would be required, inch by inch, with spotters, signals, and timing. A transformer has no patience. Steel has no mercy for a bad angle. One mistake at a hairpin, one misjudged tyre line, and the consequences would travel faster than the vehicle ever could.

The work had not started yet. The first questions had. *How would a trailer take that corner without scraping rock? Where would the rig pause without blocking the road? How would a loaded vehicle climb without losing grip?*

The initial site visit made clear that conventional methods wouldn't easily fit this project. The terrain forced them to reconsider everything they knew about standard substation design. On the spot, engineers started sketching rough ideas to handle these conditions. They decided quickly to go for a multi-step level Gas-Insulated Substation (GIS) setup rather than the single level GIS. Multi-Level GIS required much smaller volume of land development, shorter timeline, optimum utilization of space, this meant less disturbance to hillside while cutting and shaping.

Also, instead of creating an entirely new route for power transmission across Kishtwar hilly terrain, the team proposed substation near the tapping point of existing 400 kV line. This would limit how many new towers they had to build and how much land they would disturb. This simpler design meant fewer logistical headaches.

Office, Permissions, Ground Zero

Soon after the site visits, the team knew exactly what had to be done first. They needed a base, a practical place to manage daily operations. They found suitable space within Kishtwar town, close enough to the site yet still accessible from the main roads. It was a simple building, basic but comfortable. It provided the essentials: desks, reliable phones, computers, and meeting spaces. This temporary office quickly became busy with maps, engineering plans, and team discussions.

Alongside getting the office ready, there were multiple permissions and approvals to secure. The local administration had clear procedures, and the team had limited time before snow and monsoon arrived. If paperwork got delayed, it could hold back everything, including equipment movement and construction schedules. Abhishek and the others knew that clearances had to be tackled urgently.

So, they started meeting regularly with district officials. They openly shared the project plans, explained the benefits to local communities, and listened patiently to any concerns. By openly involving officials early, the team built a genuine understanding. Permissions began coming through at a good pace, land approvals, administrative go-aheads, and important regulatory permissions such as those under Section 164 and from the Central Electricity Regulatory Commission (CERC). Each step completed felt like another small victory.

By late August 2023, the paperwork was mostly settled, and the team could finally mark a date on their calendar, 19th September 2023. That was the day they had planned to break ground. Now, the focus was on making sure everything would run smoothly on that first day.

As September approached, a mix of excitement and nervousness settled in the field camp. Everything so far had been planning and preparation. Soon it would become real earthwork, steel structures, and hard physical effort. Abhishek and the entire Resonia team knew that once they started, there would be no turning back.

Sarthal Rd, 182204
21°C
70°F
2023-05-10(Wed) 10:38(am)

Chapter 3

Blood, Sweat, and Stakeholders

"Well, this is it," Abhishek said, looking down at the untouched ground. "After today, everything changes."

When a team knows the road ahead will be hard, the first step is to agree on how to move. That is how they began in Kishtwar: walk first, decide after.

They parked short of the gate line and went on foot. The access road climbed with tight hairpins. The ground rose fast. Slopes read between 30 and 45 degrees. The plot sat against the hillside, and space was tight.

They treated the road like a work front of its own. Each bend was checked for a long trailer. They marked a few places where a pilot vehicle could pause traffic for a minute while a rig turned. They listed bends that would need widening. Gradients were tested with the question that matters most: will a loaded truck make it up in rain as well as in dry weather? A few turns would need guided movement, so standing space for spotters was noted.

On the plot, they read the ground the same way. They traced where monsoon rain would run and where snowmelt might pool. They marked faces that would need cutting and faces that would need support. First bench levels were sketched: a main yard, a terrace for the transformer, another for the reactor, and a room for the control building. Ramps between levels had to be gentle so pickups and trailers could move without drama.

Winter was written into the notes. Local lows fall to about minus six or minus seven degrees. Concrete would need a method and timing. They logged spots that freeze early and patches that catch morning sun. Stores would sit on firm ground. A warm corner for crews would be set near the work, with hot drinks and first-aid within reach.

They closed the walk with one rule for the build: fit the hill and fit the town roads. Keep cut and fill low. Plan moves for windows between rain and snow. Build access that works in March and in January. With the photos taken, measurements recorded, and sketches filed away, the day's fieldwork was complete. What followed was the task of translating those observations into decisions that would guide the grid plan and the layout ahead.

Now, a Grid Plan That Fits the Hill

After the walk, the next move was to design that respected what they had seen. The substation had to sit inside a tight plot. So, the team chose a 400/132 kV gas-insulated substation at Kishtwar. GIS keeps equipment compact and lets the yard sit on the benches they had in mind for the main level, the transformer terrace, and the reactor terrace. It turned a narrow hillside into a workable layout.

Moving power out needed a link that matched the town roads and the slope. The team decided to construct a small line to loop into the Kishenpur–Dulhasti 400 kV double circuit rather than chase a long new line across hills of kishtwar. The tie-in was planned at about 1.4 kilometres with five towers. A short link meant fewer fronts, faster build steps, and movement windows that could sit between rain and snow.

Voltage support was built into the plan from the start. The yard includes a 420 kV bus reactor of about 125 MVAR. It sits on its own terrace in the scheme alongside the first bay set.

The layout also leaves headroom for what will come next. Provision is kept for future 765 kV and for extra 220/132 kV bays, so incoming hydro can be added in phases without tearing up finished work.

With the GIS and LILO fixed, the next question was space. Could the whole scheme sit on this hill and run safely? External proposals were reviewed first. Each one solved a slice and lifted the cost through larger platforms, longer internal roads, and extra retaining. The team brought the work back to their own desk.

Engineering, led by **Sandip Maity**, reopened the layout from first principles. The GIS block was turned to run with the contours. Bay spacing was reset so conductors, gantries, and maintenance paths sat on benches that could actually be cut on this ground. Terrace sizes were brought to what the equipment required rather than what a template drawing assumed. Internal roads were redrawn with modest gradients so pickups and trailers could move between levels in wet months as well as in dry.

Civil design mapped levels in detail. The yard sits about fifty-four metres above the approach road, so a single ramp would not serve. Short passing bays were added, turning pockets were placed at decision points, and a pull-in marked near the gate line for escorted moves. Catch drains were laid along the upslope and tied into cross-drains, with outfalls set where the ground could take the flow without scouring new benches. Silt traps and rip-rap were shown where runoff would be stronger in the monsoon.

Foundations for the transformer and the reactor moved to table-top concepts, so excavation stayed lean and bearing stayed uniform. Locations were checked against soil investigation notes and designed according to the relevant IS codes. Retaining lines were split into shorter runs where possible. That choice helped sequencing, kept concrete tonnage in control, and allowed crews to complete sections without holding up the next pour. Cable trenches and earthing runs were routed so they would not clash later with roads, drains, or support walls.

Quality controls were written on the drawings. Cold-weather concreting notes sat with bar schedules. Temperature logging points were fixed. Curing covers and storage pads were located on firm ground so winter nights and wet weeks would not stall progress. Stacking areas for steel, insulators, and panels were shown away from soft patches.

Safety requirements were set in the same pass. Fixed access for work at height around the GIS building was detailed. Standing space for banksmen was marked where guided movement would be used on tight turns. On the line side, a room was reserved to rig for the stringing span that crosses about one hundred and thirty metres of level difference, so crews would not be forced into awkward angles later.

By the end of this cycle, yard, roads, drains, terraces, foundations, and service runs sat together on one coordinated set, matched to the ground the team had walked.

The drawings were ready. Now the team had to move people and loads up and down the hill every day.

The yard sits roughly 54 metres above the approach road. **Zartab Alam** walked that rise with pegs and a tape. He marked where a road could climb in short steps so trucks would not strain in rain or cold. Curves were kept broad so drivers could keep a smooth line. He set small pull-ins where two vehicles might need to pass. Near the gate, he marked a turning space so a long vehicle could line up once and go through. Along the uphill edge, he laid a drain, then linked it to cross-drains that send water to ground that will not wash out.

Inside the boundary, Zartab turned terraces into working spaces. The transformer sits on its own bench with room for lifting and for a truck to turn. The reactor has a separate bench, so work there does not block the transformer's moves. The control building faces the yard so crews can reach bays and panels on foot without a maze. Roads between levels follow the same gentle rise as the approach road, so the drive inside feels the same as outside.

Growth is built in. Pads for extra 220/132 kV bays are kept free. The edge room is saved for 765 kV readiness. The internal roads already touch these pads, which means later phases can roll in on the same route without breaking what is done.

Cold nights and a sloped site ask for firm habits from day one. **Shekhar Bhau** wrote those habits into the drawings the field teams use.

Concrete pours are planned for warmer hours. Thermometers are fixed at pours that will feel the chill first. Curing steps sit next to those notes so crews know what to cover and for how long. Stacking areas for steel, insulators, and panels sit on firm ground near the work fronts, not on soft spots that sink after rain. Before each foundation pour, a soil bearing check sits on the task list and ties back to the bench name used by design and civil teams.

Shekhar also marked places to watch the structure as the hill is cut and loaded. Readings on retaining and foundation elements follow a set routine while benches open and equipment arrives. Along the benches and cut faces, he added protection lines, with drains and surface cover, so the slope holds through monsoon and through freeze. With this pack in hand, crews on site, civil teams, and engineers read the same sheet when the first trucks climb the hill.

Safety work began before the first pour. **Gulam Tahir** led it with the site leads.

For work at height inside the substation, there were places where a man basket could not be used. The plan set fixed ladders, anchor points, a full-body harness, tag lines, and a named spotter. A short rescue drill sat on the same page as the gear list and the names of trained people.

On the line, one span crosses a level change close to **130 metres**. The method showed rigging points, tested winches, pull limits, and marked standing places for banksmen. Weather holds were listed for wind, rain, and snow. Each pull started only after a quick briefing. One person gave the go call and closed the window if conditions turned.

Movements on the hill needed care. Landslip zones were mapped. Wet patches and early ice were flagged. Before a haul, the team walked the route, checked the pull-in spots, and set spotters. A phone chain was tested on calm days so people knew who to call if a load had to stop or turn back.

These controls sat inside the issued drawings and method notes. They were part of the build, not an add-on.

Having substation near to the existing **400 kV** double circuit meant fewer new wayleave asks than a long new line. Fewer new corridors meant faster permissions and fewer fronts to open.

The timeline now fits the hydropower stream from the Chenab basin. **Pakal Dul 1,000 MW** sits first. **Kiru 624 MW** and **Kwar 540 MW** follow as they come in. The operating target stays firm at **98 percent** or higher under the TSA.

With drawings and method notes frozen for the opening fronts, the focus moved to town.

Kishtwar needed to support the hill every day. Rooms for crews. An office that runs meetings and calls. Vehicles on call. Vendors who can bill with **GST**. District permissions lined up before the weather turns.

The story moves there next.

Chapter 4

Building the Team and the Base

Rain greeted the team the moment they arrived in Kishtwar.

The car worked its way up slow, winding curves, and the fog closed in mile after mile. Phone signal came and went, cutting out mid-call, then returning for a minute, then leaving again. After a point, they stopped checking the screen and started trusting their own sense of direction.

Rahul Pandey pulled out a small notebook, already worn at the edges from travel, and wrote what needed to get done. Find a guest house. Set up an office. Secure vehicles. Find vendors who would answer and show up. The first week had one focus. Reach the site every morning. Return safe each night. Keep the information organised, because losing track here would cost more than time.

By afternoon, the town started revealing itself in small scenes. A tea stall sending steam into the damp air at a corner. A scooter tucked under a plastic sheet that had seen better days. A lane that narrowed fast between stone walls and then disappeared. People watched the newcomers without saying much. Rahul met that watchfulness with patience. He showed his company ID, spoke through the lease terms, and kept his tone level. "We will follow the process," he said. "We will pay on time." Keys could wait. Trust could not.

Each evening, on the drive back over wet roads, they compared notes. A lane that felt fine in dry sunlight turned slick after a short shower. A gate that looked wide enough turned tight near the hinges. A room that seemed perfect failed on parking. A pickup truck needed a straight approach and a clean exit, no tight turns, no guessing. The notebook filled line by line, detail by detail, each correction setting up the next morning's run.

The door-to-door rounds began. Rahul carried three things everywhere: a company ID, a sample lease, and a small checklist for owners, which said, "PAN, Aadhaar, bank details for rent, and property papers." He explained why each item mattered, where copies would be kept, and who could see them. Owners listened, some with interest, some with tight faces. Many talks ended when papers came up. The fear was common. Once documents leave the hand, life becomes a chase. Rahul did not push. He answered what he could and left space for people to think.

There were pulls from the side lanes. Local fixers praised certain houses that suited their circle. Approaches were tight and turns were blind, yet the pitch stayed the same. "It will work." Rahul stayed patient and firm. Parking space mattered. Gate width mattered. A straight line from road to door mattered. He paced the approach, counted steps, drew a quick sketch in the notebook, and saw the move in his head: a loaded pickup turning in rain. If the turn failed in that mental run, the team thanked the owner and moved ahead. Care at the start would save long arguments later.

Some visits turned slowly in their favour. An owner who stood with folded arms asked to read the lease again. Another wanted to know how deposits would be tracked. Rahul showed the receipt book with serial numbers and the contact sheet with names and phone numbers. The houses they finally booked had two things in common. A pickup could enter in one go. A vehicle could wait without blocking a neighbour. Clarity eased the worry. Trust followed that clarity.

With beds set and keys working the same way each night, the team looked for a room that could hold a workday. The office they chose sat off a narrow lane. Power stayed steady most days, and the router kept a call. Two desks faced the sockets to keep cables tidy. A notice board went up on day one with the numbers that mattered: district contacts, vendor heads, drivers, site leads, and emergency services. Rahul walked the lane again with a pickup, tested the gate, chalked a turning point on the road, and marked a pull-in spot for days when two vehicles reached together. The room felt less like a rental and more like a base.

Daily calls took a fixed time. Even when the network dipped, people knew when to try again. The order stayed the same. First movements and supplies, then local shop stock and road news, then anything extra. A kettle sat in a corner with a box of teabags. Wet afternoons ran long, and the team learned the town's clock. Heavy rain pulled shutters down early. Deliveries worked best in the late morning window. Paperwork moved fastest after lunch. The board was filled with new numbers and small notes, and each scribble took weight off someone's head.

Small fixes helped every day. Plastic mats near the door kept mud off the floor. A shelf held parcels, so packets did not sink under files. A white folder named Central Papers lived on the main desk and nowhere else. Copies of key documents stayed there, so no one had to call the camp for basic checks. None of this looked big. It worked quietly and kept the day from fraying.

With lights, files, and calls running in a steady rhythm, the focus turned to the town's shops and carriers. Stores had what the site needed. The snag came at billing. Many counters gave slips that worked for local trade but failed company rules. Rahul went shop to shop with a sample invoice and a short note on the fields to fill. Some owners listened with crossed arms. Some frowned and tried one bill. When that first bill cleared, the second came easier. Shops that kept to the format joined a small list and began to see repeat orders. Payments landed on time, and calls were returned. That simple reliability opened more doors than any long speech.

Vehicles ran through vetted vendors. Local drivers asked for direct placement. The team stayed with one channel, so documents and insurance lived in one file, and one number solved a missed slot. At the front desk, a register tracked reporting and release times. Fitness, insurance, license, and registration sat together in the same drawer. Fights were reduced because the routine stayed the same for everyone. A late arrival met a time written in ink, not a debate.

Vendors started to adjust. A hardware store that shrugged at GST fields kept a pre-printed format near the counter within a week. A taxi operator who resisted paperwork returned with a full set after two weeks of steady runs went to a rival. The town learned the team's way, and the team learned the town's pace. Afternoon calls dropped. Old clarifications stopped repeating. Beds were in place. The office held firm on wet days. Shops and vehicles moved on time. The base felt real, built on clear steps and small wins, ready to carry the hill work without fuss.

By next week, the notice board will have done one more job. It pulled a small line to the office door. A mason came in first, then a driver looking for steady trips, then two men who said they could work at height. The desk kept to one routine. Hear the request, note skills, check IDs, and pass the name to the team that was hiring. If an EPC partner needed hands, we sent them there with a slip. If a function in our office was hiring, we walked the person to that desk. A single register tracked entries, and what happened next. Rahul kept a pen tied to the book so no one had to hunt for it.

The rule was plain. No false promises. If there was an opening, we said so. If there were a test, we shared what it would include. If there was nothing that day, the name still went in with a phone number. People did not like hearing "wait," yet they liked a clear path. Word spread that the gate spoke one language to everyone. A welder came because a friend told him, "They listen, they write your name, and they call if there is work." That trust helped more than any banner.

Once the yard opened, Gulam Tahir gathered everyone for a five-minute talk each morning. He stood near the tools, hands on a small checklist. The points were simple. Where we could not use a man basket in substation areas, fixed ladders and anchor points were set. Harness on, chin strap tight, one named spotter who did nothing else. No one argued with that. The crew had seen what a small slip can do on wet steel.

For the line spanned with a level change near one hundred and thirty metres, Gulam brought a printed method sheet. It showed rigging points that had been checked, the winches that had been tested, and the standing places marked with paint. Weather holds were written down, not left to mood. Each pull began only after that short talk. One person, and only one, gave the go call. That voice stayed the same for the whole pull so no one second-guessed a shout on the wind.

Haul days started with a route sheet. Landslip patches were marked, wet turns were circled, and rest points were named by distance, not by guess. Drivers knew where they could pause without blocking a bend. If the rain picked up, the sheet had a line for where the convoy would stop and wait. This looked like extra work when the sun was out. On a wet afternoon, it saved a full day.

Winter came in slowly and then settled hard. Nights touched minus six, sometimes minus seven. Concrete does not forgive in that cold. Shekhar Bhau moved to the warmer hours and fixed small thermometers at the places that caught the chill first. He kept curing covers ready on site, not in a far store. He would tap the covers and say, "These save us from coming back with hammers." No one argued with that either.

Steel, insulators, and panels sat on stacking pads on firm ground close to the fronts. Slopes tempt you to drop material where you stand. Shekhar kept the pads where the earth held and where a small crane could reach. Soil bearing checks were done and written before each foundation pour. If the numbers looked weak, they fixed the base before they mixed a bag. Readings were taken on retaining and foundation elements as benches opened and loads came up the hill. The notes were dull to read, yet they kept men off rework shifts in the dark.

Small habits stacked into a way of working. A foreman would check a ladder rung with his boot before stepping up. A helper would call out if a harness tag was faded. A driver would point at a wet patch and say, "We take the higher side." None of this made a poster. It kept people steady.

The office work ran in the same spirit. District meetings were set before the heavy rains, so routine permissions did not stall at the wrong time. The white folder on the main desk held copies of the central papers we reached for again and again: transmission licence, TSA, Section 164, and the trackers for PTCC and Defense Aviation. If someone needed to show a reference, the folder opened, the page came out, and the file went back to its place. New staff learned that on day one.

A date went onto the calendar in thick ink: 19 September 2023. It sat there as a mark to work toward. No one made a speech. People knew what had to move before that week. By early 2024, the LILO stretch showed its shape. Four of the five tower foundations were in progress. The PEB and CRB fronts kept moving. The reports came in with a small line that mattered to all of us: zero lost-time injury on those packages. It did not mean the hill had turned kind. It meant the base could take weight.

These were not grand wins. A clear register at the door. Five-minute safety talks that people could repeat in their sleep. Thermometers tied to rebar where the cold bit first. A route sheet that told a driver where to wait when rain climbed the valley. Papers in one folder that anyone could fetch in ten seconds. Each small step took the sting out of the next day.

By the time the first spring sun warmed the yard, the place felt ready for the longer push. The town knew our way of working. Vendors kept the right bills at the counter. Drivers arrived on time and left on time. New workers came with their IDs in hand because someone told them it mattered here. The hill did not get easier, yet our feet stood firmer.

Design That Followed the Hill

Space inside the substation land stayed tight from the first drawing. The plot had to hold foundations, equipment, roads and walls in a narrow frame. As work began, the engineering team realised that some of the planned foundations were landing on top of each other. On a flat site that might be solved with simple re-marking. Here, the hill did not give such freedom.

The answer came through composite foundations and cantilever arrangements. Where there was no direct space under a pedestal, the team used combinations that carried the load safely into the ground without demanding a wide footprint. Main foundations and retaining walls around the yard followed this approach. It was not a single master stroke. It was many small tweaks that respected what the land could carry.

Cable routing carried its own problem. The original plan called for nearly two thousand metres of open cable trench. Time and space did not support that. Even if every other task finished, the last stretches of trench would still be under work when the rest of the yard stood ready.

So the team changed course. Instead of long open trenches, they moved to HDPE pipe runs. Crews laid pipe, pulled cable through it and sealed it. Work became quicker and safer. Less excavation, less open trench to protect in rain and winter. Pipe runs turned into a quiet game changer for the schedule.

Tertiary foundations raised another question. Some went ten to twelve metres below ground in filled soil zones. The depth looked heavy for that kind of fill. Rather than guess, the team called in a plate load test agency from Delhi. Contracts moved at high speed. Every stage of the trailer's journey was tracked until the equipment reached site.

After compaction and testing, the results showed that the soil could support the loads with less depth than first marked. That allowed the team to reduce foundation depth where it was safe to do so. Less digging in unstable fill, less time, and a design still backed by data, not comfort alone.

Day after day, drawings changed to match what the hill allowed. Every alternate day, as he says, a new proposal went from site to engineering and came back with a revision that helped progress.

> ***Began with a wet road and a short list had grown into a base that could hold, built by people who chose care over noise.***

Chapter 5

Land and Local Dialogue

In the Morning, the Gate Stayed Shut

It was their first day on the plot after the payments cleared. When the team arrived, they found they were not alone.

A small crowd had gathered at the entrance. Women stood together, wrapped tight in woollen shawls, breath showing in the cold morning air. Men leaned against the gate with arms folded, watching without expression. A short distance away, a few boys sat on motorbikes, relaxed in posture but alert, eyes moving between the team and the road behind them.

The gate stayed shut, firm and unmoving, holding everyone on the outside for a moment longer.

Rahul Pandey, Site HR and Admin, stepped forward with his ID and a thin folder. Ankit Singh from the land side stood with him. They greeted the elders first. Names were exchanged. People spoke of jobs, past dealings, and what work would mean for the lane and the fields nearby. Rahul opened his notebook. He wrote what each person said, point by point, so no one felt skipped.

Gulam Tahir, Chief Manager HSE, asked the crew to hold back the vehicles. No engines started. No equipment moved. The team shifted to the roadside shade and kept the talk there. It felt slower than everyone wanted, yet safer for that morning.

By midday, a police team arrived with little fuss. They kept their distance and set up a small post near the sharp bend in the road, present without getting in the way. What began as a few hours turned into days, then weeks. Their job was simple: keep the situation from boiling over so conversations could happen without fear of a sudden clash.

People came and went through the day. Some watched with curiosity. Some stayed guarded. A few carried hope they chose not to name. For some, the tone changed when they saw Rahul's notebook. He wrote down what was said, line by line, and read names back with care. It sent a message that mattered. Their words had a place to land.

Others moved straight to practical questions. "What comes first?" "Whose land gets touched first?" "Who gets called for work?" They wanted to understand how this project would enter their lives, when activity would reach their fields, when jobs would open up, and when the work would reach their part of the road.

Rabindra Sahu from project execution stood close to the book and marked tasks that could begin once the gate opened. He spoke of survey marks, of laying safety tape, and of the first small steps that make a site real without noise. Heads nodded, then voices rose again. The day held like that, slow and firm, with the gate still shut and the notebook filling up. What went on the next line would decide which move came first.

"हमें जॉब दो."

The words came every day at the office door. Young men from the lanes. Older hands who knew concrete and steel. Rahul Pandey, Site HR and Admin, kept a register on the desk. He wrote names and skills. He asked for ID copies. He promised a call when a slot opened. The rule stayed the same for all who came in.

"आप हमारे एरिया में काम कर रहे हो. आपको तो हमें वर्क देना पड़ेगा."

That was the mood when the first contractor packed up. They had walked the ground for weeks, measured, held meetings, and then said the hill would cost time and money. They left. A month went by with them. The police post at the bend stayed. The gate stood watched while the company looked again for a crew that could work this slope.

The next team arrived without fuss. They met the land cell in town. Ankit Singh sat with the minutes from the gate talks. Drawings marked during those days were spread on the table. What can begin inside the fence when the gate opens? What can start outside without trouble? Who calls whom at six in the morning? A short plan took shape.

Rabindra Sahu from project execution walked the yard line with the new foreman. He pointed to small, safe starts. Mark survey points. Set safety tape on the edges. Fix a store where the ground stays firm after rain. Gulam Tahir from HSE stood with his notebook and added checks from that morning's look. The new crew kept their list tight and said they would stay. People at the gate watched and weighed those words.

At the desk, the queue kept moving. Rahul asked each visitor what they could do and wrote it down. The talk in the lane turned from anger to terms. Local hands would come in where the day allowed. Teams with hill experience would work the tasks that needed speed and method. Roughly half and half. The register began to show that balance.

Hareem Kripak walked the tight turns with site leads and showed why two spotters help when a long vehicle turns. Zartab Alam pointed to the patch where the access road would rise later. Seeing the line on the ground helped people understand how movement would work when bigger loads came through.

By week's end, there were fewer raised voices at the door and more dates in the margin beside each name. The next pages in the register would decide who walked in first when work inside the fence began.

Papers, Rooms, Trust

Rahul Pandey carried a thin folder and a small tape. He moved from door to door with a company ID on the first page and a sample lease behind it.

At each table, he read the same list. PAN. Aadhaar. Bank details. A copy of the property paper. He said why each line mattered for rent, payments, and audit. Owners listened. Some brought a tin box and set out files. Others asked for a day to speak with family.

He wrote names and times in a ruled notebook. Tomorrow at ten. Evening after prayer. He left a stamped receipt wherever a token changed hands. Copies went into plastic sleeves with the house name in pencil so the file could travel even if staff rotated.

Outside, he checked what the rooms could not tell. Gate width with the tape. A chalk mark where a pickup could swing once and leave. A corner where a visiting truck could wait without blocking the lane. If these checks failed, he thanked the owner and moved on. Better to avoid a dispute later than win an argument that slows work.

By late afternoon, he had two shortlists. Rooms that worked for people. Rooms that worked for vehicles. The circle in the notebook only came when both lists matched.

Land Follow-Ups Without Drama

Ankit Singh joined the rounds as the land face in the field. He stood in courtyards with owners and wrote plain notes. Boundary stones, a path used by a neighbour, and rainwater that cuts through a side strip in August. Each line carried a date and the name of the person to meet next.

He returned when he said he would. One point closed, then the second, then the small ones that cause friction if left open. Owners saw their words read back from the notebook. That eased the next conversation.

Back at the office, Rahul matched field work with paper. Stamped copies in one drawer. Originals in another. A wall tracker with green ticks for houses ready to sign, dots for those that needed one more visit. Beside each line sat a backup name, so files moved even when someone was away.

By week's end, the board held two marks that changed the week. A guest house with parking that a pickup could take in one go. An office with backup power and a router that held a call in the rain. The next address on the list was under a paperclip.

Late afternoon left the lane open. School vans had passed. Shopkeepers had time to talk. Hareem Kripak walked with the site leads, a short tape in his palm and chalk in his pocket. At the first hairpin, he measured from tyre mark to wall, then drew a white arc where the turn would ease if a hand's width of tar came off. He wrote the bend number in his book and noted the trim, no more than needed.

A few steps on, he paced out a pull-in the length of a small car. A pilot could wait there, slow the line for a minute, then wave a long vehicle through. The next bend had a river-stone wall on the outside with two stacks of crates. He spoke with the shop owner and pointed to a chalk line that kept the wall safe while giving a trailer one clean sweep. One crate's width moved by a foot would do. The owner looked at the mark and agreed to shift the stack when the crew arrived.

Hareem kept moving up the street. Chalk. Measure. Note. Each corner earned a line in the margin. One-minute hold here. Mirror spot there. A small board that says slow at the school gate on haul days. He marked where a scooter stand could move by a metre so the view stayed open when a rig turned in.

At the last turn before the gate, Zartab Alam joined. He stood on the high side and traced with his boot the path that would become the access road. The climb would rise in short stages and meet the internal levels without jumps that jar loads or feet. He pointed to the place for the transformer terrace, then to the reactor terrace, then to the walkway the control room staff would use in the rain. He linked those points to the pull-ins Hareem had set on the lane, so the move from town road to yard felt like one plan carried through.

People from two nearby shops stepped closer. They asked where deliveries would halt in busy hours. Zartab pressed a small square into the dust with his toe and said a van could stand there while a trailer passed. He showed how a shortcut of tar at the bend would keep a door from scraping a wall. The tea seller nodded. The tailor across the lane said he would keep his bench inside during the first few hauls.

Gulam Tahir came behind with a ruled pad. He circled a damp hollow that turns slick when rain sits. He drew a cross under the walnut tree, the patch that freezes early in January. Beside each mark, he wrote a plain rule. Pilot here. Spotter there. Hold movement after this hour in frost. He added the name of the supervisor who would take that call, so the decision had an owner. Two cones and a whistle went on the list for hand signals when the lane grew busy. These notes would sit in the method sheet so crews work the same way on quiet days and on tense days.

By evening, the chalk arcs made a pale map on the road. Children hopped over them on their way home. The constable from the bend asked where he should stand when the first trailer came. Hareem opened his book and drew a small square by the school gate, then wrote a time window that kept clear of prayer and closing hour. The lane now held a plan people could live with, and it sat in ink as well as in chalk.

What Unblocks The Gate

Talks moved from the entrance to a small room off the lane. Ankit Singh placed the minutes on the table and read them in order. Owners came in, one by one. Boundary stone at the mango tree. A cart path that cuts the corner and needs a plank on market days. Rainwater from the upper field that runs through a yard in August needs a side channel. Each line earned a tick when closed. Where a visit was needed, he wrote a date and the name of the person who would go.

Outside, the post at the bend stayed. Inside, the office phone rang when the signal held and brought the same two questions. When do we start? Who steps in first? Answers began as actions. A survey crew set pegs along the fence and called out the numbers so the elders could hear them. Yellow tape ran where the first bench would open. A store tent went up on firm ground that Rahul Pandey had booked with the owner the week before. A vendor dropped cones and rope and left a clean bill with his stamp. The pilot car took its place at bend one and raised a palm for a minute while a pickup rolled through.

Near the gate, a small crowd gathered. An excavator eased up the lane and stopped before the chain. The operator looked to the foreman's hand and waited, engine low. The register on the desk had ticks next to the promises from day one. Hareem checked the chalk at the first corner. Zartab slid the access sketch into his pad. Gulam ran a finger down the time window he had set for this slope and this season.

> ***The morning ahead would choose which line of work to begin the cut.***

Chapter 6

Making Ground from Slope to Benches

"Rain writes the first line of a workday here."

Monsoon in Kishtwar leaves evidence before anyone speaks. Tyre prints soften into wet earth. Steam curls off a battered tea pot placed a few steps beyond the bend. That is how the day opens.

The crew reaches site before dawn. Headlamps cut through mist. Jackets stay pulled tight against cold and damp. Inside one pocket, a folded list sits close to the body, the first checks for the morning: begin the initial cut, keep the side drain open, spread gravel where boots sink, then get back before the afternoon storm lets loose.

At the edge of the plot, the excavator coughs, settles, then comes alive. Its first bite goes in slow and deliberate. A watchman notes the time in a ruled register, treating minutes like inventory. Site leads take position on higher ground and point toward a faint line marked the previous evening with paint and string. The first bench stays small by intent. Safety first, progress second. A strip wide enough for a pickup to pass in one clean move. A dry patch set aside for stacking earth so rain does not throw it back across the track.

June and July bring rain that keeps returning. A morning plan can dissolve by noon. Mud climbs to boot tops. Tyres spin where soil turns into heavy paste. Two hands open the side drain with shovels, giving water a path out. Stones go into a shallow pit to trap silt before it reaches the lane below. Another pair spreads gravel where the ground refuses to hold. The foreman checks the fresh cut with a measuring staff and calls the moment the face begins to lose strength. Engines cut. Work pauses. The team waits for the earth to settle and take shape again.

Between town and site, the road changes its mood after every downpour. A morning that starts dry can end with stones piled at the hairpin near the school. The pilot car calls in from that bend and asks the convoy to hold. Two trucks pull into a lay-by the team picked out days earlier, firm ground with enough width to wait without blocking the run. Drivers knock mud off boots, make short calls home, then sit tight until the message comes through.

Back at the office, the whiteboard keeps the day in view in three neat lines: Bend three held at 10:40. Lay-by one in use. Check again after one. Every note stays logged. Every movement stays planned. In this weather, order is a working tool, same as a shovel or a radio.

By July the work learns a pace the hill accepts. The excavator cuts in short runs, then rests while water leaves the trench. The stockpile grows on the dry patch set for it. A store tent on firm ground holds rope, covers, spare boots and a kettle for crews who end up on the other side of a sudden block. A clerk pins a photo beside the plan so the next shift sees what changed in the day.

August feels heavier. Rain sits longer on the face. The access path needs small mends in the same spots. A mason places short logs at the bench edge where the track softens and lays a plank where a footbridge helps. The drain receives a second branch to a safe outlet so the lane stays clean. A simple rule goes on the wall. When rain hardens, hold at the marked bends and call in fifteen. Names sit beside those bends so a voice answer when the phone rings.

Foundations wait for two green signs. The ground under a footing must pass the field test. The sky must give a dry window long enough for the cure to take. Early September offers that window. The side drain beside the bench runs free. Gravel lies where feet once sank. The lane is passable to the yard. A return hour is set before the clouds build again over the ridge.

Cement bags arrive under cover. Water for mixing is measured and noted. A mason ties a nylon string to set the top level of the blinding layer. When the drum turns, calls are short and clear. The mix meets the pit. A small board on bricks shows the start time. Another board records the temperature and the cover to keep through the night. A rope goes round the new work with a red tag that says keep off. The notebook on the chair takes one more line. First bench cut. First foundation poured. Cure watch on. Sun feels like a visitor in July; in September it begins to stay.

By late afternoon the light stays long enough for paper. A blue tarp and two stools become a desk. **Rabindra Sahu**, Substation Electrical and Civil, spreads the sheets and sets a spanner on each corner so wind does not lift them. A senior stands beside him. They check every sheet against the hill in front of them.

Marks from the day guide their eyes. A footing that sits on a doubtful slope gets a red circle and a short note in the margin: move one metre upslope, keep the top level as planned. A cable route that meets soft ground gets a gentler turn sketched in pencil with the new distance added. Where a deep pit would waste time, the spot is tagged for a table-top base and the soil reading from the log is copied beside it.

The marked bundle goes with the driver in a plastic tube. A fresh set returns in a folder with today's date on the corner. Older sheets are stamped and filed in a crate under the tarp so crews do not mix versions after a wet night.

As the light thins, foremen gather. Rabindra keeps it practical. Start at seven at bench two. Keep this path open for the mixer. Test this face before the crane moves in. Names are paired with tasks and the sketch that shows where to begin, so the morning opens without guesswork.

The hill edits drawings in its own way. A footing marked deep on paper meets hard truth on a bench that sits higher than expected. In those spots the team moves to a table-top base so the load spreads wide without digging a deep pit that would slump after rain. Heights are set against the day's soil reading and the bench name, so the concrete crew knows where the surface must land.

Cable routes learn the ground as well. A trench that would run straight across a damp patch is turned early, then brought back to its goal along firmer soil. Bends are gentle so a pull will not snag later. Where the trench passes close to a future road edge, the path shifts a little so tyres will not crush it in a busy week.

None of this lives only in memory. The site diary holds photos with a scale, a one-sentence reason, and the name of the person who spotted the issue first. A junior can open that diary on a wet morning and see why a base moved by a metre, or why a cable route bends wide by the old walnut stump. The next hand works with that context instead of guessing.

The GIS building brought a different kind of pressure. The structure needed work at nearly twenty-five metres height. Under normal conditions, a manlift would carry workers safely up and down. The route to this site made that impossible. The terrain and the narrow climb meant the machine could not be brought in.

The team did not wait for equipment that would never reach the hill. They adjusted their methods. Crane placement was checked daily. Grounds were examined each morning before lifting began. People worked with platforms set up on site and moved carefully through each stage.

For almost two to three months, they reported early and went back late. Everyone knew how this building affected the schedule. GIS installation could not begin until the PEB was ready. The team kept at it without complaint. When the roof closed and the walls stood firm, it felt earned in every sense. It showed what steady effort can look like when nothing around you is simple.

Access That Climbs In Short Steps

From the town road to the yard, the climb goes in small rises. **Zartab Alam** marks the path with chalk dots and paint where needed. Passing spots sit where a long vehicle will need a pause. Single-turn pockets sit at bends that cannot take a two-point move. Water has its own plan too: a drain runs along the uphill edge and slips into side channels that send flow to ground that will not wash out.

Inside the fence, the same thought runs through. The road meets the transformer terrace without a jolt. It rolls on to the reactor terrace, then to the control room door. No harsh ramps that would make a mixer stall or a crew trip in the rain. Surface is rolled firm where cranes will stand, and the edges are kept neat so stores do not creep into the wheel path.

Drivers test the run in steps. First, in an empty pickup. Next, with a light load to feel how the tyres bite after a shower. Then with the long trailer, that will matter. They report what scraped, what squealed, and where a mirror needs a hand.

Changes are small and fast: a stone pared back at a wall, a marker moved by a foot, a pull-in widened by the span of a palm. By week's end, the route reads as one road from the town bend to the yard, and the people who use it can tell you where to slow, where to hold, and where a turn goes through in one clean move.

October brings a thinner sun and hard mornings. **Shekhar Bhau** sets concrete work for the warmer hours, when the mix bonds well, and men can use bare hands for finishing. A small thermometer hangs beside the bay that cools first. Covers lie next to the mixer so the slab can be wrapped the moment finishing ends. Steel, panels, and insulators stand on firm pads near the work face so they do not sink after rain. Before any batch leaves the drum, the ground under each foundation is checked with a field tool and signed off. Retaining lines and foundation blocks have set times for checks while terraces open and lorries bring loads. The same figures mean the same thing to every foreman, which keeps talk short and decisions steady.

The list that guides this sits beside the work in a plastic sleeve. Each task has a name beside it. If the wind builds or the sun fades behind clouds, the window closes and the cover goes on. No calls across the yard, no end-of-day disputes, only the next small step taken in order.

High work needs its own order. Gulam Tahir keeps one kit for bay areas where a man basket cannot reach. Ladder tied at the foot and held at the top. Fixed tie points in place before anyone climbs. Full-body harness clipped in. Tag line in hand. One named spotter on the ground with a whistle and a clean view. The same sequence runs each time a person leaves the slab.

Out on the transmission span the ground drops close to **130 metres** between towers. That pull has a clear plan. Rigging points are set and marked. Winches run with a trial load before the real run. Stands for the signal men are painted and numbered so crews use the same names for the same places. A five-minute talk sets roles. One person gives the start signal. The same person ends the window if wind builds or rain sets in. Every hand knows where to stand and what to watch while the wire moves across the valley.

Hauls on the town road follow a one-page sheet. Wet turns are circled from the monsoon months. Soft patches carry a note with the hour when they heal. A phone number at the top reaches the person who can hold a convoy or send it through. The pilot car takes the first corner and waits in the pull-in marked earlier in chalk. Drivers keep that sheet on the dash and speak the same names for the same bends, so no one is guessing at dusk.

Work moved because many people minded small things. A crane wastes time if it hangs over wet slab, so **Vikash Patel** stood with the lift plan and walked the ground with the foreman. Boom length set. Sling checked. Lift taken in the hour the slab could take it. Hooks did not hang in air.

Dates lived on a wall chart. **Niman Kumar** kept those boxes honest. If rain moved a task, he marked a new window and called the two people who needed to hear it. Hand-offs sat beside tasks, so one crew finished, the next walked in, and no one arrived to a cold front.

Small delays eat a site. **Baleshwar Mahto** cleared them with the foremen. A stack left in a wheel path. A pump too far from the drum. A shuttering board short by an inch. He moved the stack, dragged the pump, cut a board that fit, and told the lead what changed so the next slab did not wait.

Materials have their own timing in a hill town. **Ramveer Singh** matched deliveries to the hours the road allowed. He called from the first bend and asked the pilot car to hold a minute when school vans passed. A bag due at noon arrived at two, and the reason went in the log with the driver's number so tomorrow did not begin in guesswork.

Paper can slow a day as much as rain. **Rahul Pandey** watched that line. Vehicle logs by hour. Drivers who pick up on call. Bills with GST fields where they belong. If a stamp was missing, he reached the right desk in ten minutes, not ten calls. Vendors who met the bar came back again, which kept fuel, food, and spares from turning into long stories.

This was never only five names. Riggers set lines and took the strain when the word came. Banksmen held ground at the stands and gave the hand signal that mattered. Masons kept levels true. Steel fixers tied rebar with fingers cut from wire ends. Carpenters closed the formwork tightly. Survey hands carried rods and checked marks after each burst of rain. Electricians ran a temporary light to the far edge before dusk. Welders stitched brackets where the drawing asked for more hold. The batching hand kept the drum honest. Drivers held gears low on wet climbs. Spotters stood at the pull-ins. Guards kept a book at the gate and matched faces to badges. The cook in the store tent poured tea that warmed hands at four. The cleaner kept mud from drowning the office floor, so the wall chart stayed readable. Police at the bend and the district patrols gave the town a sense that work had order. Many hands made a steady day.

By early 2024, the signs showed. Four of five tower foundations on the short link were in progress.

The pre-engineered building and the control room block met their next marks with no lost-time case on those packages. The board in the office shifted from guesses to time windows. 10.30 to 12.00 crane. 13.30 to 15.00 concrete. 16.00 checks. Crews began each morning with three short lines on a card by the door: ground firm, gear ready, people named. At dusk, a single line in the log told what held and what would open first after sunrise.

> ***One window on that board stayed bright: the day the wire would run the span.***

Chapter 7

Winter, News, and Night Watches

"Air is the first tool."

On these hills, everyone learns to use it well. Work took the same pace.

Cold makes people slow down and choose each move. The team met near the gate and set the day in one small circle. Layers on. Inner wool. Shirt. Fleece. Jacket. Cap pulled low. Socks checked for dry toes. Boots tight. Harness and helmet in hand.

A board stood on two bricks. One line stayed at the top in chalk. Everyone returns safely. Below is the plan for the morning. Work fronts listed. Pairings for height work read aloud, so no one climbed alone. The person running the crane showed hand signals once, then once again. The first-aid kit location was pointed out. A small card with emergency numbers hung where a torch could find it before sunrise.

Tools were named before they moved. Ladders are tied both at the ground and at the steel. Anchor points were tried with a tug. Tag lines coiled and clipped. Radios tested with a short call and answer. One phone sat high in the office where the signal held best. If a bay floor felt greasy, the order moved, not the standard.

Down the lane, the guard post sat under a tin sheet. A ruled book lay open. Names and hours were written with a stiff pen. The district sentries noted their time, the torch, the whistle, and the radio battery. A constable watched school vans take the bend and waved them through with the ease of someone who knew the town's clock. A patrol from the SOG passed before daylight and again after dark. That sight calmed the road.

Shelter tents stood close to the work. Tea, first-aid, dry socks, and spare gloves were kept there, not far away in the store. Before the first lift, hands and shoulders went through a short warm-up. Calls were short and easy to hear. Left. Hold. Cut. Lower.

Ground and sky had the last word. If frost sat in the shade near the walnut tree, grit went down. If a step felt slick, a plank bridged it. One person set the time windows for heavy moves and said them twice. The guard at the bend heard the same times, so lane and yard kept time together.

Breathe steady. Gloves on. The hour struck on the watch at the gate, and the day opened.

News Reaches the Yard

The tea stall kept a small TV on a hook. News rolled with a line of text that moved fast. A map, some grainy clips, studio voices raised and low again. Two or three men watched with cups in their hands, then looked up the lane toward the gate.

At the yard, calls went out to homes. A mother in Kulgam. A brother in Doda. A friend in Jammu. Words were plain. Reached site. Working. Will call again. Now and then, a sentence is cut off. The phone fell silent, and the screen showed one bar that came and went. No one pressed the redial again and again. A short message would go when the line returned.

One handset stayed on a high shelf near the door. It had a long charger and a spare battery taped to the cord. That corner held the signal more often than the rest of the room. The rule was simple. One person used it, sent a line to the family list, then placed it back on the shelf.

Beside the door hung a card with numbers. Police post at the bend. District control room. Hospital. Site lead. HSE lead. The card sat in a plastic sleeve so rain could not soak it when the door swung open. People knew to point to it when a visitor asked who to call.

If the post at the bend sent word to pause, work paused. A hand rose from the guard hut, or a short call came on the radio. Machines idled. Keys went to the lock box. Crews stepped to the safe side of the line. A supervisor wrote the time in a ruled book and walked to the front. The plan at the board moved to the later slot. A runner told each team where to stand. No speeches. No rumours. Men sat on a low wall or stood with cup and biscuit and spoke of small things until the call returned.

The stall owner lowered the volume when workers came by. He said the last headline, then passed a kettle and a plate of buns. A constable from the bend took a cup too. He knew the drivers by face now. He told them when he would wave them through. The town had its own way of keeping the hours straight.

When the post said go, the board at the gate got new times. The person at the shelf phone sent another line to the list at home. Safe. Back to work. The crews took their places, and the day found its pace again. The next story the TV would carry felt far away and close at the same time, but the only thing that mattered in that minute was the hand on the radio saying one word. Proceed.

Rain hit hard across the valley. Word of a cloudburst spread before noon. Calls stopped within the hour. Screens showed a bar, then nothing. The team moved to what they could see and do.

The day's plan went on paper. Names of fronts, small tasks under each, and the order to try them in. A runner carried that page from the yard to the tower crew, then to the gang cutting the next terrace, then back again with short notes in the margin.

Open work came first. Pumps were set where water pools fast. Planks went down, so feet did not sink. Covers pulled over cable trenches. Stacks moved to higher patches so muddy flow would not spoil them. Drains cut earlier in the season were opened with shovels. Where the slope sent a stream across a path, a side channel took it to ground that could take the load. Nothing fancy. Only the steps that stop a day from falling apart when phones go quiet.

Deliveries were moved to hours when the road could carry them. Drivers who knew the bends agreed to come before dawn or after tea. A pilot car waited at the first turn and slowed a corner for a minute so a van could pass. Goods already in town were kept in a spare room that had been booked for weeks like this. No one promised a noon drop when the hill said two o'clock would be wiser.

With calls down, people used signals they all knew. A raised arm across the lane meant stop. A slow circle of the hand meant to start the pump. An open palm meant to wait until the water fell. By mid-afternoon, the runner could point to three small wins: a trench kept dry, a store tent saved from a leak, a path made firm with gravel, and a plank. By dusk, the guard at the gate read a ruled book and said each name in turn. Every person had checked in; with the time they left for rooms or stayed for the next round.

Night brought headlights on the lane. Police jeeps rolled past at set hours. SOG teams drove through after ten. Guards walked extra rounds from gate to corner and back, noting their times in the same book with a blunt pencil. When the last vehicle was filled and there were no more seats, crews went in small groups, as agreed in winter. Four to a group. One torch held high. One whistle in a pocket. The route kept to houses with lights on and shopfronts that face the road. No field cuts. No guessing at forks.

The walk had its own pace. Boots on wet tar. Breathe in short clouds. A few words between people who had stood on the same ledge that afternoon. At the bend, a constable raised his hand, counted heads, and pointed to the stretch ahead. The guest house door opened for the first group, then the next. Back at the gate, the guard ran a finger down the last page and gave the only update anyone wanted. All in.

Cold meant the body needed time. Before the first lift, hands and shoulders went through a short warm-up. Squats, arm rolls, neck turns. Crews stood in pairs and copied the lead. No rush. The idea was simple. Warm up, then work.

Shelter tents stood close to the front. Inside sat a flask, a tin of tea, a box with plasters and bandages, a basket of dry socks, and a stack of spare gloves. Stock was checked at the end of each day and topped up at first light. People used what they needed and returned what could be reused. The tent flap faced away from the wind, so the floor stayed dry.

Short lines helped when days ran long. "Hold here." "Two hands on." "Stop and step back." These calls stayed the same from crew to crew. The same words worked on a fine morning and on a hard evening. A small slip of paper with these lines lived in the tent pocket and in the site leader's jacket, so new hands learned fast.

Work also kept time with the town. School exam dates are on a printed sheet next to the haul plan. On those mornings, heavy movement paused when students walked in and out. Prayer hours were marked too. Noisy tasks shifted to the other side of the day, so the lane stayed calm when it mattered. A short list of vendors who delivered during festival weeks hung by the phone. Those names got called first. It saved a long talk later.

Rain changed roads more than any note could. After a hard spell, a slide on the approach road could stop travel without warning. When that happened, convoys waited at bends the team had already chosen in the dry months. The pilot car stood at the first corner and signalled when to move. One caller opened the time slot for travel. The same caller closed it if rain built again. That way, crews heard one voice and carried one plan. A van went through when it was safe to go through. A truck waited when it needed to wait.

By the time lights came on in the town, the tents were tidied, socks and gloves were counted, and the sheet of dates on the wall had a fresh tick for the day. The crews knew what the night would bring and what the morning would ask for.

A small card sat near the door, taped to the wall. Thick marker said, "Build a node that serves this valley through winter."

People reached for the edge as they passed. Some read the line under their breath. Others touched the tape and moved on. It sat at eye level, so no one had to search for it. The corners wore down in the cold. At week's end, someone wiped it clean and pressed the tape firmly again.

The guard at the gate knew the sentence by heart. He said it to new hands on day one, then showed the way to the shelter tent and the first-aid box. When visitors asked what the team worked toward, he pointed to the card and let the words stand.

Hard mornings made the card feel close. A rigger looked at it before climbing. A carpenter checked the last nail on a shutter after reading it once more. A driver set the time with the pilot car and glanced at the line before starting the climb. In the tent, the cook read it and poured tea a little sooner when fingers turned stiff. No speeches. A habit that kept everyone facing the same way.

At the month's end, the card turned over. The back held names in pen. Not a roll of stars. A circle of many hands. A mason who held a level on the soil that wanted to give. A signal man who kept a crane honest with one palm and a whistle. A store hand who found rope when the rain came early. A spotter who stood at the bend so a truck could pass once and pass clean. A guard who matched faces to badges in poor light. A driver who parked by the school gate and waited for a nod. Each name took a single line.

News still moved through phones. Flood weeks still bent plans. Patrols still passed the lane after dark. The card did not fix those things. It steadied the room when they arrived.

People read the line, took a slow breath, and stepped out.

Chapter 8

Money, Materials, and the Long Week

Water finds a path through rock.
Cash and supplies needed paths like that, too.

The finance desk kept a weekly sheet on a low stand by the door. Three marks told the state of play. A green tick meant money received. An amber dot meant on the way. A red circle meant a delay had reached a work front.

Friday before lunch, two people stood there. One reads bank credits and payment proofs. One marked the sheet. Field leads came in, looked once, and knew what the next seven days could hold. If a red circle appeared, the person on that task stepped across, asked which item was held up, and moved a lift or a pour to a later hour. If a green tick appeared, a call went to the transporter who knew the bends, and a time was fixed for the drop.

Trigam 05/24/2025, 15:40

Names and long totals stayed in the ledger. The doorway showed only the marks people needed to act. **Nobendu Kumar Nath** checked releases against next week's tasks so crews were not set up to wait. **Pallav Kumar Mishra** matched vendor terms with what the bank had cleared, so a promise on paper matched a truck on the road.

By two in the afternoon, the sheet was ready for the week ahead. When the rain changed the roads, the desk rang the transporter and wrote a new time in the corner. When a payment landed late in the day, someone came back and added the green tick so Saturday work could move without a round of calls.

Beside the stand sat a sample invoice under a clear cover. It showed every line the job would check. GST number as registered. Vendor name as on the certificate. Item, unit, quantity, rate, tax, and total. Clean lines. No gaps.

Bills arrived and were checked against that page. If a tax line was missing, the vendor filled it at the desk. If a digit in the GST number was off, the desk called and corrected it before filing. If two copies looked the same, the clerk looked for a unique number or a delivery note before passing it on. **Rahul Pandey** kept a thin folder of good examples so new vendors could see what was needed without a long talk.

Fridays carried a short round of checks. Stamps in the right box. Dates that matched the goods-in note. Names that matched bank details. Small fixes were made the same day, not saved for month-end. When a file was opened later, the page inside would match the work outside. The week stayed tidy, and the hill did not lose hours to paper that should have been right the first time.

On a slope, gear has to be light in the hand and quick to use. Hoses were kept in short lengths, five to ten metres, so one person could carry and lay them without tripping. Joiners sat in a small tin so two pieces could click together when a longer run was needed. Each hose had a colour ring and a tag with its length. After use, it went back as a neat coil on a hook at waist height, not across a doorway.

Covers were chosen for speed. Heavy sheets waste people and time. The store kept light tarps with stitched loops at the corners. Pre-cut ropes sat in a bundle with simple knots ready. One person could pull a cover over a small stack and tie two points in a minute. Each cover had a painted mark for the bay or block it served, which saved a search on wet mornings. When the day was dry, covers were spread on the fence to air and then folded tight so they would not flap loose later.

Electrical items were checked before they met the yard. They were opened at a clean table, the label was read, the cable and plug were looked over, and a small buzzer told if the circuit held. A hand meter gave a short insulation check. The piece then ran for a few minutes on the temporary supply to see if heat or smell warned of trouble. A green sticker showed the date and the initials of the person who checked it. Anything doubtful went on a "hold" shelf with a slip that said what failed and who would sort it.

No tool left the store without a name. The issue card had four lines only: who took it, which work front, time out, and time due back. Batteries went out, were charged, and came back to the same tray. Bits and blades lived in labelled boxes, not a heap. When a tool returned, the store hand wiped it, looked for cracks or frayed cords, and filed the card. If a tool stayed out past its mark, the store called the front and wrote a new hour. This kept the slope moving and kept the gear ready for the next pair of hands.

Some days, the hill closes the main hairpin without warning. Loose rock comes down after rain. The call comes from the pilot car at the first bend. Turn the load. Hold the rest.

Drops move to the next safe window. Times are picked that do not clash with school and market hours. The transporter hears the new slot and repeats it back. The site notes which item waits, which crew depends on it, and who will receive it when it arrives.

Goods already in town do not roam. A spare room exists for days like this. Pallets sit on bricks, not the floor. Covers are dry and tied. Small fans run for an hour to take out the damp. A slip on the door lists what is inside and the phone number of the person who holds the key.

A second transporter stays on call for dawn runs through the longer road. The route is slower, yet open. The desk checks fuel and driver rest before agreeing on the start time. The gate pass is prepared the evening before. The pilot car waits at the outer bend at first light with cones and a rope. If the post on the lane wants a pause for school vans, the van waits and then goes when the hand lifts.

Crews that expected the delivery switch tasks. Steel fixers prep ties. Carpenters cut shutter pieces that fit the next pour. Electricians test temporary lights on the far side of the yard. No one stands around guessing. When the road clears, the call goes out once. The lorry climbs in, the store hand takes count against the slip, and the item finds its mark without a second trip through town.

A table stood near the window. The pay sheet, a cash box, a phone for transfers, and a ledger. Names were read in order. Each person stepped up, said their front, and saw hours matched to what the bench log showed for that week. Where rain and slides slowed the road, the note beside a name said why a day went missing, and when that day would be made good. Landslip risk and slippery movement were not words on a form here; they were part of the week on this hill, and they were recorded that way, so no one lost pay to a blocked turn or a frozen patch in the yard.

Cash and transfers were both used. A slip stayed with the cash. A message confirmed a transfer. The line at the bottom of the page showed that the totals agreed before anyone left the room. If the list carried a new hand from that morning's hiring, Rahul Pandey's desk added the required papers first, then the name found its place in the pay line. The same desk kept vendors straight on GST fields, since more than one shop in town did not issue a proper bill at the start and needed the sample to set it right.

Small errors can drain a month, so they were taken care of the day they appeared. A missing stamp on a goods-in note was fixed that afternoon, not at the month's end. A wrong unit on an invoice was corrected before the file moved forward. A tool hired for three days returned on day three, with the store slip signed at both ends. Each Wednesday, the store hand and a foreman walked the racks. They checked coils, bits, blades, and cords, and wrote what needed work or change. Cold-weather habits from quality control sat beside this routine, because a hill in winter punishes loose ends: covers near the pour, material stacking on firm ground, and checks that hold when the temperature drops.

Where a supplier hesitated on clean paperwork, the office showed the sample invoice and the firm's list of fields. It was needed for the audit, and it was needed so the payment landed on time. Rahul had learned this in his first weeks here, when many vendors would not part with ID or bank details, and many shops did not issue GST bills. Making the process visible was how those ties formed and stayed steady.

When the window outside filled with fog or hail, the table near the window still worked through its line. A man from the stringing crew signed, a mason from the terrace front signed, and an electrician from the room test signed. People left with money in hand or with a message on their phone, and a short note in the ledger marked where the day had been hard and why. Sandip's team had kept the layout tight to this ground, and the same spirit guided the books: fit the need, remove what wastes time, record what matters.

Rain changes the day before anyone speaks. Fuel and food were planned to miss the school rush and the market hours. Morning drops came early, before uniforms filled the lane. Evening drops came after shops pulled shutters. The desk kept a small list of local alerts and matched runs to those times so a van met an open road, not a crowd.

Signal fell away without warning on wet days. A phone chain bridged the gaps. One call went to the transporter. That person called the driver. The driver called the pilot car at the first turn. If a leg failed, the link above made the next call. Messages were short and clear. "Leaving now." "Hold ten minutes." "Passing the bend."

At the bend, one person stood with cones and a rope when rain raised the risk. A minute of still traffic let the van climb without horns or tight squeezes. Drivers who knew the hill took a lower gear, waved a thanks, and rolled on. Inside the yard, the store hand was ready with a tally slip and a dry patch for cartons. Food went to the tent near the work, not to a far room where people lose time. Fuel met the machines in a steady order, so crews did not queue in the rain.

These small moves are tied back to the weekly sheet. Each release backed a real front. Concrete for a slab. Crane time for a lift. Gear for a pull. Spares for a check. When a tick went on the sheet, someone booked the run, and someone met it. When a dot stayed amber, a lead moved the task before workers reached an empty spot.

The yard grew because the week held together. Not from a single big push, but rather from many clean passes that did not waste hours. Before the lights went off, a short note sat under the weekly sheet. One line on what moved today. One line on what needs help at nine tomorrow. The last pen stroke set up the morning.

Work on the line brought its own mix of challenges. Some tower locations stood on land where owners had questions or concerns. Each case needed patient explaining. Teams met people on those plots, showed why the tower needed that point and listened to what they wanted addressed. Progress depended on cooperation at each location.

Access was another concern. Several stretches had no motorable path. Material had to be carried by head load across steep ground. Angles were sharp and the climbs tiring. Crews moved steel in smaller units, walked it slowly to each tower point and kept going until everything needed for that tower reached the spot.

These two parts of the job took time. The project moved because the teams stayed steady and the people along the route gave the space needed to finish each span.

Next came the days when the short link and the ready yard had to move in step.

Chapter 9

Bringing Power Through

First power is a set of small agreements kept in the same hour.

The day's map set a calm order, bay by bay. Teams knew where to start, what to check, and the short slot for going live. Roles were clear before hands touched gear.

First came the look-and-touch pass. Covers off. Doors open. A torch across joints and glands. Loose packing removed, stray tags set right, fasteners firm, gaskets seated. In GIS sections, gas pressure was read against the gauge beside the bay and noted. This made the space fit to test.

Interlocks followed in the sequence on the diagram. Keys moved in turn with spoken cues so nothing overlapped. A refusal where it must refuse is counted as correct. A freed key when the next step needed it counted as correct. Alarms were then tried: local lamp, hooter, and the desk indication answered back. Where SCADA was ready, the screen showed the same state and was read aloud.

Insulation checks came next. The five-kilovolt set measured bus sections, cable cores, and control wiring where required. Values and room temperature were called and logged. The acceptance band sat on the page, so no one hunted for a manual mid-test. Quality holds were printed in red and meant a pause until earthing resistance, housekeeping, tool clearance, and interlock logic were confirmed.

Only then did a short live window open. One feeder at a time. Close, observe, measure, and open again. Voltage present, current where expected, no stray alarm. Temperature points were scanned with IR at ten, thirty, and sixty minutes on joints, links, and terminations that reveal early warmth if contact is poor. If anything read wrong or felt wrong, the map sent the team back to the last safe step, a small fix was made, and the try was taken again.

With the day's map set, the team turned to the bays. Panels were placed in their slots and fixed true. Each panel tag matched the schedule. Temporary supply lit the facias, lamps came on, and labels read the same as the list. Field leads walked the line and called out panel numbers so crews knew where to start and where to stand.

Bus sections were then closed up. Faces were wiped, gaskets sat even, and fasteners tightened to the mark shown for that joint. Viewing windows and gauges were read, and the reading went on the bay sheet. Links and access covers were shut in order so anything from trapping a tool or a tag. The aim was a sealed path from disconnect to breaker to bus, ready for trials.

Alarms and trips ran in a set sequence. Local lamp first, then the sounder, then the indication at the desk. Where the desk showed the same state, the person at the desk said it aloud, and the bay team moved to the next point. Trip checks used the safe path chosen in the plan. Each trigger had a witness, and each result had a short note with time and bay name. If a step did not behave as drawn, the test stopped, and the team traced back to the last good point before trying again.

Before anyone spoke of first energisation, the earthing grid was checked. Continuity from each bay to the main earth bar was proved, and the contact value was noted. Bonds were looked at end-to-end. The earth pit nearest the bay was read and logged. Covers went back only when the values sat where the method sheet expected them.

The order of work followed the design lists prepared by Harish Kumar. Those lists named the panels, the checks, and the order they made sense in. Field leads turned that into work fronts people could follow. Tools set out. Access kept open for the next move. Sheets are ready for entries that would matter later. By the time the last panel of the set was tried for alarms and trips, the bay read as ready for the short-lived window the map allowed.

The bus reactor sat on its own bench above the yard. The terrace had a clear lane from the gate so a crane could reach it on any service day without disturbing other paths. Crews kept that lane free. No pallets on the curve. No stacks near the turn.

Before the first energisation, the team took a baseline. The housing was wiped clean. Nameplate details were read out and entered. A hand held a probe to set the vibration reading at three points, then a second pass confirmed the same numbers. An oil sample went into a labelled bottle and into a small case for the lab truck. A technician pressed the earth tester to the studs and spoke the contact value. The recorder wrote it beside the bench name.

The terrace had small things that made work easier. Timber blocks already set where jacks would stand if the unit ever needed lifting. A drip tray is placed under the drain to keep the surface clean during sampling. Shade cloth was tied above the work spot so people could see the instruments in the glare. A tray for tools, a bin for used wipes, and a rag line for gloves that needed to dry before the next use. The living part of the job looked after itself this way. People finished an action and put the place back in order, so the next crew did not have to hunt for basics.

When checks were complete, the cover guards went back, bolts were painted to mark their position, and the terrace log stayed with the sheet for that bay. Crane access was walked again with the civil lead, so future arrivals would not be a surprise.

Winter and the Fight for One Sound Cube

Foundation work began in late December. Benches were cut into the slope so that foundations could sit on firm steps instead of loose skin. That was when the real winter problem showed up.

Concrete that should have set in six to eight hours took almost twelve to fourteen hours. Cubes stayed weak even after curing. Every engineer on site knew that a bad foundation in this climate would become a story the hill told for years, and not in a good way.

The team ran through twenty to thirty different options. They tried different cement combinations. They changed temperatures. They varied the duration of heating and curing. Each round meant more cube samples, more tests over three days and more waiting for numbers that still refused to rise.

Slowly, a method emerged. One cement grade worked better for this site than the others. A settling time was fixed. Heating time for each foundation moved into a band of three to seven days. Once they followed that routine, the cube strengths started reaching the level the project needed.

On the ground, this method turned into tents and heaters. Crews covered foundations with sheets and tarpaulin, set heaters inside and treated each pit like a small room that had to stay warm through the night. Gravel, sand and water were kept from freezing. Even the cube tank had heating support so that test samples did not lose strength to the cold before they reached the lab.

Labourers and security guards sat in those tents through the night. Their job was simple and hard at the same time. Keep the heaters running. Watch the temperature. Call if anything went wrong. Outside, the thermometer often showed minus seven or minus eight. Inside, the foundation had to stay within a narrow range that the test sheets now demanded.

Work did not stop for snow. The pace changed, but the cycle of cutting, placing, heating and testing carried on. By the time winter loosened its grip, the team had something better than a story of survival. They had a tested process for building concrete in that valley, learnt the hard way, cube by cube.

Protection and Control That Talk to Each Other

The protection team set up in front of the relay racks. A short list sat on a clip: relay names, setting files, and the order to load them. Settings were loaded with a witness at each rack. The witness called file name, date, and checksum before a save. A printout went into a clear sleeve behind the door.

Trips were proved along the safe path first. A primary injection on low current showed the trip chain in slow motion. Lamps came on. The trip flag moved. A contact at the desk changed state. The person at the desk read the point name aloud. The person at the rack answered and moved to the next test. When that sequence held, a live trip test was done in the agreed window, with eyes on the yard and a hand near the abort switch.

Control links needed their own round. The control room and the bay had a dry run where commands were sent with no energy to the gear. Close, open, select, indicate. Each command had a sender and a receiver. Names sat beside those roles on the sheet, so a handover during a call did not lose the thread. When the dry run matched the plan, the same handshakes were taken with the system live. Short phrases kept the room steady. "Select done." "Indication received." "Step complete."

The people around this work showed the same care outside the room. Meals were timed so the bay crew and the desk could keep a window without hunger rushing them. Extra gloves and a clean cloth near the relays so smudges did not hide flags. A spare headset on the ledge in case a cable failed. Small habits made the place workable for long days.

By evening, each relay page had a tick, each trip had a line with time and witness, and each control handshake had both names written beside it. The room felt ready for the next step because every task had found its answer, and every person knew who would speak the next word when the call began.

By afternoon the yard began to live like a station. Spares sat where hands could find them. Contactors with contactors. Relays together by rating. Gaskets wrapped and labelled. Cable lugs sorted by size. The store counter gave out tools with a name and time, then took them back the same way. A drill went back to its place. A meter returned to its tray. A harness hung near the door where the next team could see it.

New records started from that hour. Bay trips went on a simple form with four lines. Time. Bay. What happened. What was done. Small fire actions used the same form. Late call-outs from the line used the same four lines. People filled them without fuss and moved on. No one needed a speech to know why it mattered. When the phone rang later in the night, the person who answered would see what had happened in the day.

Security kept its steady pace. Guards walked at set times and checked the same latches. Torches were tested before dusk and batteries sat charging near the desk. Gates were checked in pairs after dark. One guard carried a radio. The walk covered the bay doors, the fence near the lane, and the bend where the crane road met the inner road. Calm movements, nothing loud. The value came from knowing the same route would be walked again tomorrow.

Quality kept an eye on the ground that holds the place. Small paint dots on retaining lines and footings showed where to read. A number taken early in the week, another late in the week. If the figures stayed steady, the hill was doing its part. If they moved, someone looked for water pooling or a tyre mark in a soft corner and fixed it before it grew.

Service toward town mattered more now. The 132 kV feed to the district was proven in steps. A planned load went out while a team watched meters. Voltage was noted near the yard, then again near the edge of town. Protection operations were made to act in a safe way, and the time taken was written. Calls went both ways so hand-back after each test was quick and calm. The aim was simple. People on the town side should be able to plan their day with this route in mind.

Winter makes faults on the old line. This build gave an alternate path for those days. A feeder could carry supply when ice and wind made trouble on the legacy route. That single fact meant a shop could keep stock without fear, a clinic could plan a night run, a class could keep its exam hour. Crews knew it and made notes with that future in view so operators in town saw what they needed on paper and trusted the switch when they used it.

At dusk a small drill took place. Radios in hand. A call from the room to the yard. A gate opened, then shut. A short walk to a bay door and back. The station sounded different at night. People learned that sound. They learned their place in it. The aim was not to create fear. The aim was to have a habit that holds when rain or frost takes the easy choices away.

What this work gives the grid carries a number people can use. Availability stays at or above ninety-eight percent under the service frame. That mark sits in the day's talk the way a speed limit sits in a driver's mind on a narrow road. It is not for bragging. It keeps choices honest.

What this work gives the river valley carries names. Pakal Dul brings one thousand megawatt into this pool first. Kiru follows at six hundred and twenty-four. Kwar adds five hundred and forty. Space for more bays and higher voltage was set aside when the first lines were drawn, so new units can connect in stages without tearing up finished routes. The bench for the reactor can be reached by crane without crossing a live track. The lane to the control room lets a van turn without a risky reverse. Old morning decisions, returning as less trouble in new seasons.

What this work gives the town comes closer to the skin. A fault no longer steals a day. An exam holds its hour. A bakery keeps trays. The people in the yard do not call this victory. They call it a fair return on weeks that pulled hard and weeks that went by in a blur. The stringing crew who made the steep span safe used two words at the end that fit here too. Week safe.

A station runs on parts and on good timing. A fitter saw a stain near a flange and tucked cloth into a pocket before stepping out. An electrician found a warm lug on an infrared scan and planned to tighten it in the cooler hour. A planner moved a crane slot to miss school time on the lane. A driver parked a van on bricks in rain so cartons stayed dry. None of them asked for a poem. All of them asked for a clear list and a path that stayed open. The place rewarded that.

Money and supply moved with this life. Weekly marks from the desk meant something to the field. A green tick meant a drum or relay would arrive when hands were ready. An amber dot meant patience and a swap in order. A red circle meant a talk and a small change that saved a day from slipping away. Finance did not lay concrete or string wire. Finance kept hours from being wasted.

Care for people stayed built into the day. Guards knew faces. A sentry who waved a lorry in also knew how to let an ambulance through. Winter tents near the work held tea, bandage, dry socks and warm gloves. That tent served crews twice a day. No one wrote songs about it. People used it and went back to the task.

Training kept to the ground. New staff from O&M walked with a lead and took simple tasks first. Open a cabinet. Shut a cabinet. Read a meter. Call the value. Try a trip. Clear a trip. Learn to stand two steps back from a bay door.

They need repetition until they sit in the hands.

Chapter 10

A Node That Holds Through Seasons

A day holds together when the lights do.

On a full working day, this stretch of homes, shops, and small businesses pushes demand to around 145 megawatts. For years, most of that load depended on one weak link: a 132 kV transmission line carrying roughly 70 MW. Two smaller 33 kV feeders added support, pulling around 25 MW each, yet the system still ran close to its limits. People built routines around the uncertainty. Candles stayed within reach. Phone chargers got plugged in near windows where reception held. Water pumps ran at odd hours when power returned without notice. Clinics ran quick checks on backup generators before dusk, because losing supply after dark could not be a surprise.

Then Kishtwar's new yard came online, and life became easier in ways people could feel.

side, supported by a second winter route built to hold during rough storms. When ice loads the older line, or a branch drops onto a span, supply can move to the alternate path without delay. Lights stay on while crews locate the fault and clear it. The sweeping blackout that once rolled through the entire area, leaving homes and businesses stuck, now feels like an old problem.

You see the change in ordinary moments. A baker finishes a batch without the oven cutting out halfway. A welding shop completes a gate with one clean bead, no break, no restart. A classroom begins an exam on time, without students waiting for fans to start or lights to return. Refrigerators lean less on loud backup units. ATMs. and payment machines keep running through the afternoon rush. Streetlights come on when they are supposed to, instead of hours late.

A new 132 kV feed now comes in from the town Outside town, the difference shows up too. Milk coolers hold the night's yield without worry. Pharmacy fridges stay consistent, avoiding the warm-cold swings that can spoil medicine. Mobile towers keep calls from dropping across neighbouring towns at once. Fewer blackouts also mean fewer sudden resets, less damage to equipment, and fewer frayed tempers at the end of a long day.

Operators on the town side gain more control. With two **132 kV** routes, routine work can be moved into shorter windows. A line patrol can be set for a calm hour. A switch can be opened and closed once, without a rush. When weather turns rough, a message to the local team carries a time that holds, because the alternate path keeps load while the fix goes ahead.

Shops open at the time painted on their shutters. Small mills run from eight to four, not eight to whenever power returns. Clinics book procedures in daylight and keep to the slot. Students see lights stay on during study hours. The change is steady rather than loud, and it reaches every lane in simple ways.

This is the gain from a node at Kishtwar: a reliable **132 kV** route into town and a winter path ready when the old line fails. Outages ease. Repairs feel calmer. Daily plans hold.

Kishtwar serves as the pooling point for Chenab power. **Pakal Dul** comes first at **1,000 MW**. **Kiru** follows at **624 MW**. **Kwar** adds **540 MW**. The yard has room for these arrivals so running bays are not disturbed.

Space is reserved for tomorrow's needs. Pads and ducts are ready for extra **220/132 kV** bays. Road turns match those pads so trailers can reach them. Crane access to the bus-reactor bench and to future bay lines stays open, which lets a heavy unit be placed or lifted without blocking live paths. Provision for a **765 kV** link sits both in drawings and at marked points on ground, so new steel can stand on bases already cast.

Handoffs are set in advance. Times are agreed with the upstream station before a trial. The matching bay is named on the sequence sheet. Interlocks and settings are prepared in that order. When a hydro unit starts trials, its bay stands ready. When that unit enters service, the next bay in the list moves up. Control calls use short lines that crews know well: "Select done." "Indication seen." "Step complete." Power from the river reaches the node and leaves it in the same hour through equipment that has room to work.

Growth lands on prepared pads, not on new cuts. This saves time, saves money, and keeps people on forward work rather than repair.

From Test Weeks to Normal Weeks

After first power holds, days move into a routine that keeps service reliable through rain and cold. Each morning begins with a short walk of the bays. Latches, lamps, gauges and earthing points are checked in the same order. Anything odd is written once and given to a named person.

Records are read each evening. Relay events are copied, then cleared. Meter values go where the file expects them. If an alarm came and went on its own, the trace still goes in so a pattern can be found if it repeats.

Small slips are closed before noon. A warm lug found on an infrared scan is tightened when load sits low. A loose tag is replaced. A door that drags gets a clean and a light coat of grease. These small acts prevent lost hours later in the week.

Calls in the room stay short. "Select done." "Indication seen." "Step complete." The person at the desk and the person at the bay use the same lines in rain and in sun. A command lands once the previous step is confirmed. This keeps pace even when weather presses.

Tools and spares are kept where hands reach them fast. A relay on the shelf matches the rating on the panel. Gloves, visor and cloths sit at the bay entrance. Guards walk at set times and record gates and doors that matter for a live yard. Civil teams read marked points on retaining lines and footings during the week so the ground record shows how load holds across the season.

By week's end the station knows its state. Morning checks done. Repairs closed before lunch. Relay records read at dusk. Names beside each action. That is how test weeks become regular weeks, and how Kishtwar carries power through winter and monsoon while the river projects feed in, one after another.

Safety and Watches Tuned for A Live Yard

Work at height in a live yard needs care and time. Where a basket cannot reach inside a bay, teams use fixed ladders that are tied at the base and at the top. Tie points are checked before anyone climbs. A second pair of hands checks harness buckles and the karabiner. A tag line is set so a tool can be guided rather than fall. Cold mornings slow fingers, so gloves are warmed first and clips are tried twice. The same order runs each time. A line in the record notes who went up, who watched from the ground, and what task was closed before boots touched the slab again. Some days this takes longer than planned, and that is accepted, because steady steps keep people safe.

The steep span that stretched nerves during stringing sets the ground rules for upkeep. Stands for banksmen carry numbers so people reach the right spot without guesswork. Winch pads are checked before any pull. Hand signals match the sheet used during the build, and a whistle covers the turn where faces cannot see each other. Wind can rise across the drop and fog can flatten depth. The team follows a simple call in such hours. If sight fades or the pull feels uneven, make the span safe, coil lines, write the time, and return when the window improves. New hands learn beside seniors who worked the first pull. They practice on short sections before they go near the long fall. Progress comes from patience and from people who say when a step needs to wait.

A live yard also needs eyes on its edge. Guard rounds are written with time and place, then walked as written. The path touches the gate, bay doors, inner bend and fuel store. Torches are tested before dusk and spare batteries sit on charge. Police jeeps pass the lane at hours agreed with the station. SOG teams add a late pass. These visits help on festival nights or when the road carries more vans than usual. The round spots a stray dog sleeping near a wheel path, a visitor lingering by a fence, or a van parked close to a bend. A short word moves things along and the lane stays usable.

Briefings keep their five minutes even when breath steams in the air. Crews stand near the gate and hear the plan for the day. Pairs are named for height work so no one goes up alone. People point to first-aid and the nearest shelter by habit, not by memory drift. Anyone can raise a hand if a hazard was missed.

Design Lessons to Carry to The Next Hill

The ground taught the team where drawings needed to bend to reality. Where soil near the surface lacked depth, foundations took a table-top base that spread load without deep digging. That choice saved time in wet weeks and kept benches intact. The internal road works better as short climbs, not a single steep reach. Small pull-ins before tight bends let a pilot car pause traffic for a minute while a long vehicle turns once and moves on. Those pockets cost little and remove daily friction.

Cable routes stayed healthy when ducts and markers were placed on day one. With markers in place, later crews did not cut into a live path while adding a bay. Pre-cast items helped. Ready drains, small plinths and anchor blocks went in fast and cured in the yard store before rain pushed the schedule. Standard clamp kits, labelled by size, cut search time during yard work. Spare conduits were pulled to future bay lines while trenches were open, which spared new cuts months later.

Two habits from commissioning belong in the next hill yard. Keep maintenance reach in mind when placing heavy gear, so a crane can arrive without crossing a live track. Keep a short call sheet for switching and tests, so the person at the panel and the person at the bay speak the same words and close steps in the same order. These notes are simple to carry forward and save hours when weather, roads and people are already working hard.

Care in Cold, Care on Slopes

Cold months change the day. Concrete is poured in warmer hours so it sets well. Thermometers sit at corners that lose heat first. Readings are taken before a pour and again after covers go on. Covers stay near the work so the slab is wrapped as soon as finishing ends.

Ground checks come before adding any new bay. A quick field test shows what each footing can bear. If the number is low, the plan waits. Retaining walls and footings carry small paint dots where readings are taken through the season. Monday's line sits next to Friday's line, and change is picked up early. As live load builds across winter and rain, the record shows how the hill behaves.

Slopes ask for their own care. Access climbs in short steps rather than one steep reach. Grit goes down where frost holds. Drain mouths are cleared after bursts of rain so water leaves the track. Where boots slip, a plank or mesh gives grip. Tools sit close to the front so crews are not sent back in poor light. Shelter tents stand nearby with tea, first aid, dry socks and spare gloves. Five minutes there can save an hour later.

Work at height keeps to one order. Ladders are tied at base and top. Anchors are checked before climbing. Harness is clipped and checked by a second hand. A tag line guides tools. A named spotter stands below. On the steep span, marked stands for banksmen remain, winch pads are examined before any pull, and hand signals match the sheet from stringing days. If wind rises or fog lowers, lines are made safe and the team returns when the hour is kinder. Progress comes from patience and from people who say when a step needs to wait.

Finance Frame and Licence That Keep the Light On

The station runs within a steady frame. Kishtwar Transmission Ltd is the special-purpose company. Revenue follows a **35-year** term under the TSA. The transmission licence sits in file with the orders that granted it. Section **164** powers are recorded and used when line work requires them. These papers give the room to plan work and to keep service going.

Money lines were set early and back daily needs now. Debt and equity raised for the build continue to support spares, tests and upkeep. Release of funds matches live fronts so a crane slot, a relay order or a repair can land when crews are ready. When roads close or papers slow a step, tasks move to the next window rather than forcing a bad hour.

The effect is practical. People can plan a week that the bank and the contract will support. The licence sets bounds. The TSA sets the return and the service target. Section 164 enables a line job once consent and safety are in place. The rest is steady records and a team that keeps to them. That is how power stays on through winter and monsoon in a yard on a hill.

People Who Keep the Clock

A weekly plan is posted where everyone can see it. Blocks for bay checks, line work, and town-feed tests match daylight and weather. The person who owns that sheet walks through it with field leads before any mark goes on it. Calls are brief. Timings fit the day.

Supplies move during safe hours. Fuel vans avoid the school run. Food drops come before tea or after shutters come down. If rain closes the hairpin, the transporter sends a new time. The pilot car takes that slot at the bend and the store lays out a dry space so unloading never stalls a crew.

Tools leave the counter with a name and a return time. A drill goes out charged and comes back to the same tray. Bits and blades sit in labelled boxes. When a tool runs late, the store calls the work front and writes a new time. No speeches. The day stays in order.

Admin fixes small leaks before they turn into lost hours. A missing stamp on a goods-in slip is put right that afternoon. A wrong unit on a bill is corrected before it reaches a file. Pay happens at a table near a window so names are read in order and hours match the bench logs. If a road block took a day, the note records it beside the name. Cash and transfers both get a mark so no one leaves unsure.

Security watches the edges. Rounds cover the gate, bay doors, the inner bend, and the fuel store. Torches are tried before dusk; batteries stay on charge. Police drives pass on a routine both sides know. SOG adds a late pass. Faces become familiar, which helps when a van climbs in rain.

Quality tracks the ground that holds the place. Small paint dots on retaining lines and footings mark where to read. Monday's number sits beside Friday's number. If a figure drifts, someone looks for pooling water or a rut and sorts it. It reads as care rather than ceremony.

At close of day a line on the wall records what moved; the next line lists what needs help at nine tomorrow. People see their part in those two lines and leave knowing what the morning will ask.

What the Node Means Over Years

Across Doda, Kishtwar, and Padar the power cuts grow fewer. Workshops keep machines running for full working hours. Clinics book a slot and finish within that slot. Schools ring exam bells on time. Shops lock at the hour painted on the shutter because lights stayed through the evening. Farmers with milk in coolers stop losing stock to long outages.

For the wider grid, the hill now carries weight. Pakal Dul feeds in first at 1,000 MW. Kiru follows at 624 MW. Kwar adds 540 MW. Space remains for extra 220/132 kV bays and for a higher link when needed. Pads and ducts already point to where the next bay will stand, so later work does not chew up finished ground.

Money lines back the routine. Each weekly release matches a task with a name. A check. A part. A repair. Finance moves a number, a driver brings the crate, a foreman opens it at the hour set the week before. The colour on the wall sheet changes because a small job finished when it was meant to. That is how months learn to pass without drama.

Years will leave marks. Bearings need grease after long winters. Door closers tire. Paint flakes on the sunny face. The reason the place keeps going is the same as on day one. Planning keeps the calendar honest. Supplies arrive when roads allow. The store knows who is waiting for which tool. Admin cures paper snags on the day they appear. Many hands, one day that lands where it should.

Outside the fence, families lived this work in their own way. Phone screens showed missed calls and short messages, then the tower went dark. Some nights passed without news. Others shook with downpours on tin roofs and the grind of a slide on the ridge. Winter closed the pass. Summer sent cloudbursts that turned tracks into streams. There were days with convoy halts, checkpoints and curfews. A van set out and turned back because the road had gone. A driver slept in the cab beside a landslide. Parents kept milk in coolers and hoped the power would hold till morning. Children studied by torchlight and later under a bulb that came back.

This sits behind every bay check and cable pull. A call from site reached a home, a sister, a son. Messages read safe, delayed, crossing in the morning, back by dark. When the tower returned, a burst of pings filled the room and breath eased.

> ***Their days and nights give this node its meaning. The next pages carry their words.***

Conclusion

Families and the Long Year

Every evening followed the same routine.
Check the signal. Check the time. Call home.

Everyone on the project had someone waiting. A parent stayed up, listening for a ring. A partner kept dinner ready when rain stalled the convoy at a bend. A child asked the question that came up in every house, "This week or next?" The site stood on a hill because people chose to stay on, yet the reason sat elsewhere, in homes that kept running while days stretched longer than planned.

Winter made the work harder. Temperatures fell to minus six, sometimes minus seven. Fingers stiffened. Tools felt harder to grip. Then there were weeks when the valley got cut off. Floodwater tore through, phone lines went down, and two days passed with no calls. Word spread fast, and the phones stayed silent. Patrol jeeps started moving more often as border news grew tense. Guards added extra rounds. Weather forced decisions again and again, with landslides, roadblocks, and routes that changed overnight. Many days started before dawn and ended late.

Homes learnt to live with that rhythm. Birthdays happened over video calls that froze mid-sentence. Exam dates stayed marked on calendars, with a quiet condition attached to them, only if travel allowed. Parents asked for one last call before sleep. Partners checked travel times and passed them along before dinner. Small routines held everything together. A lamp left on near the gate. A phone kept charging by the window where signal came and went. A bag packed and waiting for an early start.

These months were made of small things. Messages sent when calls dropped. A neighbour's knock when there had been silence for too long. Shoes left at the door for a late return. A note passed along after a van made it past a slippery hairpin. A quick word at the gate on cold mornings so nobody headed out alone.

Calls carried the day back home. When rain broke the signal, they tried again. During the flood, there was no choice, only waiting. When the lines returned, updates stayed short because they had to. Date. Time. Safe.

At night, the news played low on televisions. Patrol jeeps moved along the lane. Families began to recognise bends by name and learnt which hours were safest for travel. Many held their phones at the window, waiting for a single bar.

Weather kept pressing. Landslides dropped stones across narrow roads. Drivers waited in lay-bys until pilot vehicles waved them through. Dinner moved later. Sleep came in pieces. A partner kept rice warm on low heat. A parent looked out for headlights. A child fell asleep with the phone close by, hoping to hear it ring.

In time, home became the anchor. Exams stayed marked on calendars. Medicine stayed cold through the night. Shops started working on reliable hours. Families mentioned these changes in their calls without making a speech out of it. Streetlights that came on when they should. A freezer that held its cold. Milk that lasted until morning.

In those small updates, you could hear the work on the hill travel far beyond the site. It reached kitchens, living rooms, and lanes many kilometres away. It stopped being only towers and lines. It became daily life.

They said this

- **Gulam Tahir, Chief Manager HSE:** They said they worry when they see local news. They also said the company is employee friendly and takes care of people.
- **Rahul Pandey, Site HR & Admin:** They said the "war season" weeks felt hard at home. During floods the network went down for two days. Calls failed. Relatives started checking around. A short "all safe" helped the house relax.
- **Rabindra Sahu, Substation Electrical & Civil, Project Execution:** They said he is the only child and far from home. Long hours are hard for health. They listed what they watch from afar: quake risk, harsh weather, nearby attack reports, landslides after rain, food breaks, power cuts, network loss. Later, after he sent photos of the yard, they said they felt proud.
- **Shekhar Bhau, Project Quality Assurance:** They said steep ground and cold weeks worried them. They also said they were proud that he worked on an important national asset and praised his effort.
- **Sandip Maity, Head of Engineering:** They said they were excited and proud when he showed the land profile and the planned build. Early doubts were real. Soil holding and roads inside the yard were concerns. He explained the fixes and they said they felt peace of mind.

This work carries fingerprints.
Not the kind you see, but the kind that stay behind in decisions made early, calls held late, and problems met face to face. Every name here marks a hand that steadied the work when it mattered.

Jammu and Kashmir Administration

- **Manoj Sinha,** Lieutenant Governor UT of Jammu and Kashmir – supported the development projects benefitting people of J&K.
- **H Rajesh Pradesh,** Principal Secretary, Power Development Department – supported project by helping in resolution of issues and ROW guidelines.
- **Suresh Chander,** Additional Secretary PDD, OSD to Principal Secretary, Power Development Department – helped project team in resolution of right of way and project issues.
- **Ramesh Kumar,** Divisional Commissioner Jammu – helped project team in approvals and right of way issues.
- **Devansh Yadav,** Deputy Commissioner Kishtwar- guided and helped project team in revenue and right of way matters.
- **Rajesh Shravan,** Deputy Commissioner Kishtwar- guided and helped project team in revenue and right of way matters.
- **Pankaj Kumar Sharma,** Deputy Commissioner Kishtwar- guided and helped project team in revenue and right of way matters.
- **Khalil Poswal,** Senior Superintendent of Police, Kishtwar – helped project team by providing police protection and guidance on security related matters.
- **Irshad Ahmad,** Deputy Superintendent of Police, Kishtwar - helped project team by providing police protection and guidance on security related matters.
- **Nirbhay Sharma,** Tehsildar, Kishtwar – helped and guided project team in land revenue and right of way matters.

NHPC Leadership Team

- **Suresh Kumar, GM, NHPC (Kishtwar),** Provided steadfast support to the project during various critical situations in Kishtwar, ensuring smooth progress amidst challenges. His guidance and timely interventions were instrumental in resolving key on-ground issues.
- **Vinod Kumar, GM Electrical, NHPC,** Extended crucial support during emergencies by facilitating critical materials and machinery required for uninterrupted execution. His responsiveness and technical backing significantly strengthened project resilience.

CVPPL Leadership Team

- **Ramesh Mukhiya, MD, CVPPL,** Offered unwavering support throughout the project tenure, enabling seamless coordination and sustained momentum. His leadership played a vital role in steering the project towards successful outcomes.
- **Amrik Singh, GM, CVPPL,** Actively contributed through regular reviews and effective resolution of key issues impacting project progress. His proactive approach ensured timely decision-making and smooth execution at every stage.

Resonia Leadership Team

- **Pratik Agarwal,** Managing Director, Resonia, provided strategic leadership and set the vision for the team.
- **Arun Sharma,** Chief Executive Officer, provided strategic leadership and decisive oversight, enabling effective execution and governance
- **Prashant Sinha,** Chief Operation Officer - enabled effective project execution by providing timely guidance, approvals, and decisive leadership
- **Ruhie Pande,** Group Chief Human Resource Officer, defines and governs the HR policies and frameworks across the organization.
- **TAN Reddy,** Head Regulatory and Policy-provided strategic leadership on policy and regulatory matters, enabling informed decision-making and regulatory compliance.
- **Rajji George,** Director – Corporate Finance, provided strategic financial oversight and governance, strengthening capital planning, financial discipline.
- **Pooja Khanna,** Provided financial stewardship and governance that ensured fiscal discipline, effective risk management, and sustained stakeholder confidence throughout the execution of the Kishtwar Transmission Project.
- **Pankaj Kumar,** Executive Director – KWTL, led endtoend delivery of the Kishtwar Transmission Project, integrating engineering, logistics, and stakeholder coordination across extremely challenging Himalayan terrain.
- **Stanislaus Fernandes,** Assistant Vice President – Project Cost Excellence, provided outstanding contribution in project cost governance and financial control, enabling cost discipline, informed decisionmaking, and effective budget management throughout the execution of the Kishtwar Transmission Project.
- **Divyanshu Asthana,** Head HR – Resonia, aligned people policies and incentives to the execution realities of complex project delivery.
- **Sandip Maity,** Head of Engineering, provided decisive engineering leadership and design oversight for safe and constructible solutions.
- **Kamlesh Garg,** Head Land- resolved substation land acquisition and compensation challenges to enable timely execution readiness.
- **Sachin Khandelwal,** Vice President – Corporate Finance, led financial and regulatory coordination supporting COD, cashflow discipline, and closure.
- **Pawan Kumar Singh,** Vice President – Contracts & Commercials, strengthened vendor alignment, and commercial risk control.
- **Nitesh Ranjan,** Vice President – Operations & Maintenance, ensured operational readiness and regulatory compliance for smooth COD and stable operations at Kishtwar.

Resonia Project Team

- **Abhishek Garg**, Assistant Vice President, stayed close to the push when days grew tight and pressure climbed.
- **Gulam Tahir**, Chief Manager HSE, put rules in place that held through rain, cold, and fatigue.
- **Shekhar Bhau**, Project Quality Assurance, tied winter work to checks meant to last beyond the season.
- **Hareem Kripak**, Transmission Line and ROW, walked bends, marked pull-ins, and gave the span its working rhythm.
- **Vikash Patel**, Substation Erection, timed lifts so cranes moved with purpose and never stood idle.
- **Rabindra Sahu**, Substation Electrical and Civil, Project Execution, turned drawings into routes crews could follow on the ground.
- **Ramveer Singh**, Supply Chain, found workable hours along roads that refused easy passage.
- **Rahul Pandey**, Site HR and Admin, kept rooms, vendors, and records steady when phones went silent.
- **Niman Kumar**, Planning, linked one task to the next and kept the chain intact.
- **Baleshwar Mahto**, Construction, cleared snags before they grew into lost days.

Resonia Cross Functional Team

- **Meghavi Singh,** Chief Manager – HRBP, anchored people processes and HR partnering to the realities of remote, highrisk project execution.
- **Zartab Alam,** Lead – Substation Civil Engineering, translated complex civil designs into constructible, safe, and executionready substation solutions.
- **Harish Kumar**, Lead – Substation Engineering, provided steady technical leadership across OEM coordination, BOQ finalization, and engineering closure.
- **Nobendu Kumar Nath**, Chief Manager – Finance (Projects), ensured financial rigor through funding, BGs, amendments, and reporting—keeping execution unimpeded.
- **Pallav Kumar Mishra**, Finance, kept Fridays honest so the following week could begin clean.
- **Rajesh Chauhan**, Planning Team, offered second eyes where precision mattered most.
- **Ankit Singh**, Assistant Vice President – Land Acquisition, led land and stakeholder coordination, resolving ROW and compensation challenges critical to timely execution
- **Ankur Vijay,** Chief Manager – Project Finance, strengthened project delivery through financial analysis, funding coordination, and disciplined cost control.

- **Gaurav Kumar,** Manager – O&M (Substation), ensured safe commissioning, charging clearances, and operational readiness of the Kishtwar substation
- **Kumar Gaurav,** Assistant Vice President – PMO & Cost Excellence (O&M), reinforced cost discipline, planning controls, and operational performance.

Resonia Partners

- **Srikanth Iyengar,** Senior Vice President – Toshiba, provided strategic leadership and OEM partnership support, enabling advanced GIS and transmission solutions for the.
- **Biswa Ranjan Senapati,** Head – Electrical & Execution, Toshiba, led EPC execution and commissioning of large EHV/GIS substations, strengthening OEM delivery for.
- **Ravi Teja**, Toshiba, led EPC execution and site coordination for EHV/GIS substation projects, contributing to timely delivery and commissioning support
- **Devendra Pimpalkar,** Promoter Shreyas, led execution of substation civil works at Kishtwar, including slope protection and associated construction activities.
- **Firdosh Alam,** Shreyas, led onsite execution and coordination of substation civil works, contributing to steady construction progress and issue resolution at Kishtwar.
- **Neeraj Choudhary,** Director – NS Infra, led execution of transmission line civil and erection works at Kishtwar.

This work stands because these hands stayed on it.

Every name carries a home behind it. Families learnt new timings and held faith in a short call on hard days. Power stayed longer in homes. Exams began on time. Milk kept through the night. Shops closed when they said they would. The work now lives in these small gains.

We end with the words that began many calls. *"How are you today."*

Whatever the reply, the crew will take stock, pick the next task, and move it forward.

Chapter-wise Summary

Introduction

The book opens with a routine question exchanged at site, "How are you today," and uses it to frame the lived reality of the Kishtwar project. From the outset, the work is positioned as Resonia's responsibility on difficult ground, shaped by terrain, weather, access limits, and safety decisions taken daily. The introduction establishes that progress depended on consistent routines carried by engineers, workers, planners, administrators, and families waiting far away.

Chapter 1: Why This Project?

This chapter explains why Resonia took on the Kishtwar transmission project and why the location matters for Jammu & Kashmir. It connects hydropower capacity in the Chenab basin with the need for dependable evacuation and routing. The chapter references projects such as Pakal Dul, Kiru, and Kwar, and presents Kishtwar as a pooling and transfer point designed for long-term service. The purpose is framed through outcomes that affect daily life, including improved supply reliability and reduced disruption during faults.

Chapter 2 – From Bid to Breakthrough

This chapter traces how Resonia moved from bid evaluation to project ownership. It records early constraints identified by the team: slopes exceeding 30 degrees, extreme winter temperatures, narrow access roads, logistics risk, and local sensitivities. A critical decision is documented here: locating the substation close to an existing 400 kV double-circuit line, limiting the new transmission stretch to roughly 1.4 km with five towers. The chapter also covers SPV acquisition, initial financial planning, and mobilisation of Resonia's core execution team.

Chapter 3 – Blood, Sweat, and Stakeholders

This chapter shows how Resonia's design teams adapted plans to match ground reality. It explains the choice of a compact 400/132 kV GIS configuration, multi-level terracing, and a short LILO connection instead of a longer corridor. Civil planning receives focused attention, including retaining structures, drainage paths, and internal access. Safety provisions, such as fixed access points and restricted lifting zones, were built into design decisions rather than treated as site adjustments.

Chapter 4: Building the Team and the Base

This chapter shifts to Resonia's operational setup in Kishtwar town. It covers office establishment, accommodation, vehicles, vendor onboarding, and document systems required to support hill execution. Practical details such as access widths, parking constraints, billing formats, and communication routines are highlighted. In parallel, site designs evolved through composite foundations, reduced foundation depths supported by plate load tests, and the replacement of open cable trenches with HDPE pipes to suit terrain and climate.

Chapter 5: Land and Local Dialogue

This chapter documents the first major challenge faced by Resonia on ground: restricted access despite completed payments. Local employment demands and uncertainty led to gate-level standoffs. The chapter records how structured dialogue, administration support, and police coordination helped reopen access. Labour registers, repeated engagement, and transparent processes eased tension. It also describes how vehicle movement was planned through marked paths, turning checks, and controlled entry schedules.

Chapter 6: Making Ground

This chapter captures Resonia's execution phase under difficult site conditions. Terracing, bench formation, foundations, and drainage progressed under monsoon pressure and unstable soil. Work sequencing shifted frequently based on rainfall and slope response. A significant constraint is recorded here: erection of the pre-engineered building without a manlift due to access limitations, which required alternate methods and longer durations. Civil works and tower foundations advanced alongside continuous design coordination.

Chapter 7: Winter, News, and Night Watches

This chapter details Resonia's winter execution strategy. Sub-zero temperatures affected concreting, lifting, and worker movement, leading to stricter work controls and safety checks. Security conditions and communication failures added pressure, requiring simplified reporting systems and fixed contact protocols. Coordination with authorities and disciplined movement planning helped maintain continuity during a volatile period.

Chapter 8: Money, Materials, and the Long Week

This chapter explains how Resonia managed finance and supply flow on a remote hill site. Weekly tracking aligned fund release with actual site needs. Store systems controlled tool issue and return, while buffer locations were used when roads closed. Wage payments followed verified attendance and work logs. The chapter shows how aligning procurement with site capacity reduced idle time and material loss.

Chapter 9: Bringing Power Through

This chapter covers commissioning under Resonia's supervision. It outlines inspection routines, GIS pressure monitoring, interlock checks, SCADA integration, insulation resistance testing, and earthing verification. Winter concrete failures led to revised curing practices using heaters, protective covers, warmed aggregates and water, heated cube tanks, and continuous temperature recording. Energisation followed a step-by-step sequence with monitoring after each stage.

Chapter 10: A Node That Holds Through Seasons

This chapter reflects on outcomes after commissioning. The Kishtwar node strengthened supply reliability by offering alternate routing during faults and winter stress. It supported evacuation from nearby hydropower projects and reduced outage impact across the district. The chapter also notes future readiness through planned space provisions and infrastructure allowances, along with Resonia's operating routines for inspections and maintenance.

Conclusion: Families and the Long Year

The conclusion returns to the people behind the work. It reflects on long deployments, brief calls home, and families managing uncertainty while teams operated in difficult conditions. The project's value is measured through continuity of power and safe return of personnel rather than milestones alone. The book closes by acknowledging Resonia's collective effort across design, execution, administration, and on-ground support.

External Stakeholder Management

How the Ground Was Kept Open

Work in the Chenab Valley does not move on drawings alone. It moves when people understand what is coming, when officials know the order of work, and when issues are raised early enough to be settled without noise. From the first weeks in Kishtwar, the project team treated stakeholder management as part of execution, not as a parallel activity.

The district administration became a daily point of reference. Regular contact was maintained with the Deputy Commissioner's office, police leadership, revenue officials, and local field staff. Tehsildars, Naib Tehsildars, Patwaris, station house officers, and chowkilevel personnel were kept informed of what would move next, which road would be used, and where support might be needed. This steady engagement helped ensure that access, security, and administrative coordination stayed aligned with work on ground.

Issues that could slow progress were not allowed to linger. Matters related to right of way, land access, and compensation were flagged early and carried through the appropriate channels. Where local resolution required wider alignment, the team relied on established institutional forums. Constraints were raised through PRAGATI reviews, Cabinet Secretariat meetings, and coordination platforms involving the Power Development Department, NHPC, and Chenab Valley Power Projects Limited, helping align multiple agencies around shared decisions.

Compensation matters were handled with care and presence. Close coordination with the district revenue machinery ensured assessments were timely and disbursements followed due process. Fieldlevel engagement reduced uncertainty for landowners and prevented small issues from growing into disputes. When visits were honoured and payments followed records, confidence settled.

Alongside administration, the project stayed close to the community. Local employment was addressed as a working arrangement, not a promise. Registers were maintained, names recorded, and opportunities shared as work fronts opened. Conversations remained open at the gate, in offices, and along lanes near the site. Concerns were heard, written down, and revisited until they closed, allowing work to continue through difficult phases without confrontation.

In Kishtwar, stakeholder management meant showing up early, speaking plainly, and returning when one said they would. It meant using formal systems when scale demanded it and personal engagement when trust was required. By keeping administration, institutions, and local communities aligned with the rhythm of execution, the project moved forward across terrain and seasons without losing its footing.

Internal Stakeholder Management

Holding the Inside Together

In Kishtwar, internal alignment mattered as much as access on the ground. The work demanded decisions that cut across engineering, planning, safety, contracts, finance, land, and operations—often within the same day. If these functions did not move together, the terrain exposed the gap immediately.

From the outset, internal coordination was treated as part of execution, not as a reporting layer. Design teams, construction leads, planners, and support functions were brought into the same conversation early, before work fronts opened. Site realities—slope behaviour, access limits, weather windows, and local sensitivities—were fed back quickly so that drawings, schedules, procurement actions, and approvals stayed anchored to what the field could carry. Issues were surfaced as they appeared, without delay or dilution.

Daily and weekly reviews functioned as working tools rather than formal rituals. Matters discussed in one cycle were expected to show movement in the next, whether through a design refinement, a sequencing adjustment, or a commercial clarification. Engineering focused on constructability, planning remained flexible to absorb weather and access disruptions, and safety and quality were embedded into decisions rather than added later. When something could not progress, the reason was stated plainly so alternatives could be evaluated without confusion.

Clear ownership across functions helped prevent overlap and drift. Execution teams focused on delivery. Engineering ensured feasibility and closure. Planning managed dependencies and recovery logic. Commercial and finance teams ensured approvals, amendments, and payments did not stall site activity. Site administration maintained stability around people, accommodation, logistics, and local interfaces. This clarity allowed decisions to be taken quickly, even under pressure.

There were moments when priorities pulled in different directions. Speed competed with caution. Cost discipline competed with access solutions. In those moments, internal discussions stayed factual and grounded. Decisions were taken with the full picture in view—what the ground allowed, what the contract required, and what safety would accept. Once a direction was set, the team moved together, without secondguessing at the site.

Internal stakeholder management at Kishtwar was not driven by hierarchy or escalation. It was sustained through consistency. Calls were returned when promised. Actions followed discussions. Feedback from the field shaped decisions upstream. That loop helped the project remain steady through long weeks, difficult weather, and shifting constraints.

By keeping internal teams aligned to the same rhythm as the site, the project avoided fragmentation under complexity. The hill demanded coordination. The organization learned to match it.

Company Bio

Resonia Limited is a private sector power transmission infrastructure developer and solutions provider with a strong presence across India. Its portfolio includes 24 projects that are operational, flipped, under development, or under construction. Together, these assets span approximately 13,671* circuit kilometres of transmission lines and substations.

The company has built and executed a distinct asset monetisation model and has successfully monetised transmission assets worth INR 52,844 crore to date.

Resonia integrates global practices into its project execution through focused partnerships and technology adoption. The company uses aerial tools such as drones for planning, surveying, and inspection, and deploys helicopters for stringing and tower erection in challenging terrain. These methods support faster execution, improved safety standards, and consistent delivery across complex geographies.

It builds and operates transmission projects across India. Our teams plan routes, design yards, raise towers, and keep lines running through heat, rain, and winter. We work with local partners, train people on safe practice, and keep our word on timelines and payments.

The Kishtwar Transmission Project is one of our proudest efforts. Engineers, field crews, suppliers, and neighbours came together to add a strong node for the district and carry power from the Chenab basin. We measure success in steady evenings at home, clinics that can run machines through the night, and classrooms that stay lit.

This book grew from field notes, daily calls, and lessons earned on site. It was written by the project team with support from colleagues in planning, engineering, construction, quality, HSE, finance, supply chain, land, HR, and operations at Resonia.

About the Author

Pankaj Kumar has worked across power generation and transmission projects in India for more than two decades. His career has grown through years spent on site, in planning rooms, and in programme leadership roles where timelines, risk, and coordination decide outcomes.

He began his professional life in large thermal power projects, working through engineering, commissioning, and system handover. Over time, his responsibilities expanded into programme control, schedule planning, dependency management, and execution oversight across complex, multi-location projects. This progression gave him a strong grasp of how technical decisions, resource flow, and people intersect during demanding assignments.

As Project Director, Pankaj has led work that required close attention to sequencing, progress tracking, and risk visibility. His approach has been shaped by prolonged exposure to difficult terrain, tight timelines, and teams operating under pressure. Rather than relying on theory, his judgement draws from repeated engagement with site conditions, coordination challenges, and delivery constraints.

Trained as an electrical engineer and later in management, he brings together technical depth and organisational discipline. His experience spans both generation and transmission, giving him a wide view of how infrastructure systems are planned, executed, and sustained over time.

Many voices appear across these pages. Engineers, supervisors, planners, contractors, administrators, and field teams speak through their work, conversations, and actions. Pankaj brought these strands together by listening, recording, organising, and writing, while holding the full span of the project in view.

JCB

REDMI 10 PRIME | SUDHANSHU KUMA
02/12/2023 15:38

2024/8/3 18:11

2024/9/13 10:28

TEREX
TLB 740S
BPT
Trigam 03/01/2025, 16:41

Trigam 06/16/2025, 10:07

Trigam 07/24/2025, 16:12

Trigam 09/24/2025, 18:31

Trigam 09/06/2025, 09:58

Kishtwar Transmission Project:

A Journey Recognised by IPMA

The Kishtwar Transmission Project stands as a story of persistence, judgement, and disciplined execution in one of the most demanding environments in which power infrastructure is built. This journey was formally recognised when Kishtwar Transmission Limited (KwTL) received the National Project Excellence Award under the IPMA Project Excellence – Medium Projects Category at the IPMA Global Project Profession Forum 2026, held in Mahabalipuram on 13 February 2026. The award acknowledged not only what was delivered, but how it was delivered—through structured project management, innovation, and deep stakeholder engagement.

Conceived to evacuate 3,014 MW of hydropower from the Chenab Valley into India's Northern Grid, the project carried national importance from the outset. Yet its execution unfolded far from routine. The terrain of Kishtwar offered little margin for error: steep slopes, fragile geology, frequent landslides, long access chains, and winters that closed sites for months at a time. Added to this were regulatory and stakeholder complexities unique to the region. Progress depended not on scale alone, but on daily decisions taken close to the ground.

Over 36 months, the project team translated planning into action under conditions that tested both systems and resolve. Construction methods were adapted to suit terrain limits. Engineering solutions evolved when access reached its edge. Schedules were continuously recalibrated against weather windows and site realities. Through this, the project was completed within the approved cost of ₹4.07 billion, reflecting strong financial discipline and governance even as conditions shifted.

The IPMA assessment framework evaluates projects across People and Purpose, Processes and Resources, and Project Results. Kishtwar's strength lay in balance across all three. Leadership remained aligned to objectives, while empowering teams to adapt on site. Safety was treated as nonnegotiable, resulting in zero Lost Time Injuries across the project lifecycle, including during highrisk winter operations. Stakeholder engagement was sustained through dialogue rather than transaction, building continuity in a region where trust determines access as much as engineering does.

Innovation was not pursued for distinction, but out of necessity. Each constraint forced rethinking—of logistics, sequencing, and execution strategy. These choices, taken collectively, enabled continuity where conventional approaches would have stalled. The IPMA jury recognised this ability to convert constraint into structure as a defining feature of project excellence.

Beyond commissioning, the project's impact continues. It strengthens the Northern Grid, supports clean energy integration, and contributes to regional development by improving power reliability. For the organisation, the award marks a milestone—not as an endpoint, but as affirmation that disciplined project management, grounded leadership, and respect for context can deliver outcomes of global standard, even in the toughest terrain.

IPMA GLOBAL PPF 2026
MAHABALIPURAM, INDIA
PLATINUM SPONSOR
SRM
PLATINUM SPONSOR
SRM
SILVER SPONSOR
RAMAIAH UNIVERSITY OF APPLIED SCIENCES
IPMA

Recognition of the Kishtwar Project Team

Message from the Project Director, Kishtwar Transmission Limited

The successful delivery of the Kishtwar Transmission Project is, above all, a story of people—of teams who chose perseverance over comfort, discipline over shortcuts, and purpose over uncertainty. Executed in one of the most demanding environments in the country, the project reflects the very best of what Resonia stands for.

From the outset, the Kishtwar assignment tested every assumption of execution. Extreme Himalayan terrain, prolonged winters, fragile geology, access limitations, and an evolving security and regulatory landscape made progress neither predictable nor linear. Yet, through these constraints, the project team demonstrated resilience, adaptability, and professional judgement—day after day, season after season.

What distinguishes the Kishtwar team is not only that the project was delivered within approved timelines and cost parameters, but that it was achieved without compromising on safety, quality, or integrity. Teams worked through heavy snowfall, landslides, and prolonged site isolation. They remained present on ground during periods of heightened uncertainty, maintaining continuity of work while safeguarding people and assets. This quiet determination, often away from visibility, formed the backbone of project success.

The team's approach embodied Resonia's core values in action. Speed was balanced with caution. Innovation emerged from necessity rather than convenience. Respect guided engagement with local communities, authorities, and partners. Integrity remained nonnegotiable in decisionmaking, even under pressure. These principles translated into practical outcomes—adaptive construction methods, disciplined financial control, robust stakeholder engagement, and a safety culture that protected lives in highrisk conditions.

Importantly, the Kishtwar Project was never an individual effort. It was the result of close coordination across engineering, execution, commercial, finance, safety, administration, and operations teams—supported by families who bore the unseen cost of long deployments in remote locations. Each contribution, whether on site or from support functions, played a role in sustaining momentum through adversity.

As Project Director of Resonia, I extend my deepest appreciation to every member of the Kishtwar Project Team. Your commitment has delivered a project of national importance and, in doing so, strengthened the organization's confidence in its ability to take on the most complex challenges. The Kishtwar experience sets a benchmark—not only for what we build, but for how we build it.

This achievement belongs to all of you.

Pankaj Kumar
Project Director, Kishtwar Transmission Limited

Indian Member of
PMA India | IPMA® international project management association

Mahabalipuram 2026 | IPMA Global Project Profession Forum

National Project Excellence Award 2025

To

resonia

Resonia Limited

WINNER

For

**Kishtwar Transmission Limited (KwTL) Project,
a 4.07 Billion INR ($45.23 Million) transmission project aimed at evacuating hydropower from the Chenab Valley in Kishtwar to India's National Grid, strengthening national energy security.**

February 13, 2026
Mahabalipuram, Tamil Nadu

Dr. A Sivathanu Pillai
President – PMA India

Pankaj Kumar

Abhishek Garg

Akshay Dhar

Nobendu

Rahul Pandey

Rabindra Sahu

Hareem Kripak

Ramveer Singh

Gulam Tahir

Niman Sharma

Shekhar Bhau

Vikas Patel

www.ingramcontent.com/pod-product-compliance
Lightning Source LLC
LaVergne TN
LVHW070936210926
848482LV00003B/253
* 9 7 9 8 9 0 2 9 6 7 6 6 8 *